The Nature of Human Nature

The Nature of Human Nature

Dr. Carin Bondar

ISBN 978-0-557-45793-9

Illustrated by Brian C Krümm and Chris Hall, What Box Studios (www.whatbox.com)

Dedicated to those who love me and support me beyond all others, my husband, Ian, and my mum, Joanne.

Contents

Introduction

Is There Nature in Human Nature?

Are humans a part of nature? It can be argued that our species evolved just like any other and that all our actions are justified under the umbrella of what is considered to be natural. However, one could also argue that in many respects the actions of *Homo sapiens* are so far removed from those of our animal cousins that we cannot be classified in the same way. This book investigates the 'nature' in human nature by taking a closer look at some of the unnatural things we do and discussing them in the context of pertinent examples from the animal kingdom. I have repeatedly observed that issues we *Homo sapiens* face in our everyday lives are paralleled in the natural world, although the circumstances can be somewhat different. The fact of the matter is that biology is as important to the persistence of our species as it is for any other creature, and when our actions are broken down into their key components, we find that they may not be so unnatural after all.

Charles Darwin's universally recognized phrase *survival of the fittest* contains a lot more biology than one would think. To survive is paramount for all organisms. However, to be the fittest isn't exclusively to be the toughest or the strongest; it is an overall measure of survival and reproductive capacities. Biological fitness is defined as the ability of an organism both to survive and to pass on its genetic blueprints to the next generation. So in a sense, to say that only the fittest survive is a little misleading in that it leaves out direct reference to reproduction.

The most important thing to keep in mind is that the term *fitness* includes not only a survival component (how well you're equipped to survive in a certain situation) but also a reproductive component (how likely it is that your genes will be a part of the next generation). Both are of paramount importance, and one doesn't mean much without the other.

Biologists generally assume that animals behave in such a way as to maximize their fitness, so much so that the biological mantra *survive and reproduce* is central to all studies on animal ecology and behaviour. Whether you are a miniscule parasite or a massive mammal, nothing trumps the importance of maximizing your biological fitness. The fascinating biodiversity of our planet is reflected in the plethora of strategies that animals use to meet their basic needs of survival and reproduction. From the most elaborate display of beauty to the most venomous defence mechanism, planet Earth is witness to an immense collection of ways to live and breed.

Is our own species exempt from such strategies? Are the actions of *Homo sapiens* united in their goal of ensuring that we survive long enough to contribute our genes to future generations? In short, yes and no.

Although most of us manage to fulfill our biological destiny, humans spend an inordinate amount of time and energy engaging in actions that obviously do not contribute to our ability to survive or reproduce. Luckily, we are not the only species that seems to bend the rules of biology. I delight in the knowledge that inconsistencies and imperfections dot the biological landscape of our planet.

We aren't the only species that has sex merely for pleasure, and we aren't the only species that slacks off from daily chores, eats junk food, or divorces our partners—and thank goodness, because this makes us more a part of the overall biology of the planet than not. The major difference is the extent to which such things occur. While some organisms might bend the 'survive and reproduce' rules a little here or there, our species bends them completely out of shape on a daily basis.

This book is split into two sections: *Survive* and *Reproduce*. In the first section I discuss many aspects of survival that are barely recognizable as such from a human standpoint. In the western world we take many things for granted, such as the availability of food, shelter, and medical help, that most other organisms in the animal kingdom work extremely hard to attain. The irony of the situation is that, in many cases, our survival is compromised in spite of the ease with which we can meet our basic needs. The second section is concerned with the almighty process of reproduction. It isn't a simple undertaking in any species, from finding an appropriate mate to gaining the consent of said mate to producing healthy offspring. Many organisms don't get it right on their first attempt at any of these, and just like the fallible human, they've developed certain strategies for getting some practice ahead of time.

In today's world, the biology behind our actions is almost always ignored, which in my opinion is where the human animal veers from its natural trajectory. We have much to learn, not only from examining the reasons for certain actions we take, but also from observing the parallel actions of our animal cousins. It's likely you will not wonder how another organism would handle the crisis you're currently facing, but rest assured that somewhere in the world, at this very moment, another member of the animal kingdom is facing a similar challenge, whether it's about partners, children, food, jobs, or health.

This animal is likely strategizing from the point of view of maximizing its biological fitness, and maybe that's an angle we humans should take into account more often. Is eating a lot of donuts going to contribute to my survival? Will telling off my boss have a positive impact on my ability to secure resources for my family? Will making excuses for my cheating boyfriend increase my chances for reproducing successfully? If we could only reflect a little more on the biology behind our actions, the world would be a much simpler, and perhaps happier, place to live.

PART I: SURVIVE

Chapter 1

Everyone's Gotta Eat

Why Humans Need a Food Pyramid

From a very early age, humans are introduced to a balanced diet. Most of us are familiar with the food pyramid that tells the approximate percentage of fruits, veggies, dairy, grains, and meat to eat on any given day. Are we so far gone as biological entities that professional intervention is required to remind us to eat an appropriately proportioned diet?

A general survey of the food-related health ailments that decrease biological fitness in the western world indicates that yes, this kind of intervention seems to be necessary. Somewhere in our evolutionary journey, the average member of the species *Homo sapiens* has lost the propensity to select a well-balanced diet. How does our species compare to others in the animal kingdom when it comes to selecting an appropriate ratio of food items in the diet? After all, the ability to select an optimal diet is one of the most important aspects of survival in all animal species. Are we alone in our lack of ability to select a suitable balance of macronutrients and to compensate for the times when such a balance is unavailable? The answer might embarrass you, and if it doesn't, it should.

German cockroaches (*Blatella germanica*) are known as *extreme generalists*. Their diet is composed of a wide variety of foods that vary in their nutrient composition. This type of generalist strategy is akin to the omnivorous strategy adopted by our species. It is advantageous in that: 1) a greater resource base is available for generalists than for organisms specializing in one type of food; and 2) food mixing allows for an appropriate balance of nutrients to be achieved at any given time.

In an effort to determine just how well German cockroaches are able to achieve such a balance, researchers exposed juveniles for forty-eight hours to one of three diet types: a: diet deficient in protein, a diet deficient in carbohydrates, or a diet not deficient at all.[1] After the dietary regimen had been established, the cockroaches in each treatment group were given a choice of foods from all diets in order to assess whether the deficiencies could be corrected. So astute is the German cockroach that after a mere forty-eight hours in the food choice arena, all nutritional imbalances had been corrected: individuals actively replaced what had been missing in the experimental diets and then proceeded with a normal balanced diet. The accuracy of this correction is described by the authors as unprecedented among animals.

Although omnivorous animals such as cockroaches and humans can adjust the relative amounts of nutrition in their diet simply by selecting foods with different nutritional values, carnivores do not have the same luxury of choice. As obligate predators, carnivores are faced with a greater task when it comes to maintaining a balanced diet. However, organisms such as predatory beetles (*Agonum dorsale*), wolf spiders (*Pardosa prativaga*), and web-building spiders (*Stegodyphus lineatus*) are still capable of compensating for nutritional deficiencies.[2]

After a twenty-four-hour pretreatment period when they were deprived of fat or protein, the beetles and the wolf spiders demonstrated a preference for a diet that contained more of the food of which they had been deprived, successfully compensating for the dietary imbalances they had suffered. Web-building spiders were found to take the complexity of nutrient selection a step further. In the wild, these organisms are restricted to eating the prey that gets caught in their webs, and so they are not at liberty to hunt actively for other sources of food. Instead of hunting for alternative foodstuffs, web-building spiders have evolved the ability to extract specific nutrients from individual prey items. The spiders were found to extract a higher percentage of nitrogen from their prey if they had been exposed to a diet deficient in protein (which contains a high concentration of nitrogen).

Humans are remarkably less sophisticated than *Stegodyphus lineatus* when it comes to selecting an appropriate diet. Most of us are completely unaware of our nutritional deficiencies, much less how we might alter our eating habits in order to correct them. As members of our species are constantly dealing with health issues related to nutritional imbalances and poor diets, perhaps it's time we paid a little more attention to the tiny creatures around us who are knocking it out of the park when it comes to their nutrition—and they don't even have a handy pyramid diagram to show them how it's done.

Junk Food: a Human Delight!

Food choices in the animal kingdom are assumed to adhere to the tenets of *optimal foraging theory.* An animal is expected to weigh the costs and benefits of including certain foods in its diet so as to maximize survival and reproduction.[3] Whatever the specifics of a certain organism's diet might be, biologists assume that is the best diet for that species based on factors such as risk of predation, nutritional quality, and the amount of work it takes to obtain and digest it. So where do the food consumption strategies of *Homo sapiens* fit in when it comes to optimal foraging theory?

Let's discuss our obsession with junk food. Does the consumption of junk food help to maximize our chances of either survival or reproduction? Clearly the opposite is true. The negative effects of junk food on our overall health are obviously a threat to our survival. Besides that, the more pounds we pack on, the less attractive to the opposite sex we become, which ultimately has a negative effect on our reproductive output. Although humans seem to take the consumption of unhealthy food to the max, I can't help but wonder if there are other members of the animal kingdom who do the same. Are there any animal species that fill up on junk food instead of adhering to the principles of optimal foraging? The answer is maybe.

A *junk-food hypothesis*[4] has been developed to describe a situation occurring along the North Pacific Coast of North America over the last three decades. In recent years, there have been declines of up to 80per cent in populations of Steller sea lions (*Eumetopias jubatus*) and in several species of sea birds. These declines have been partly attributed to a climate shift in the ocean that has led to a complete restructuring of the fish community from high-quality, high-fat species (capelin, Pacific herring) to low-quality, low-fat species (walleye pollock). Experimental studies on captive sea lions have confirmed that it may be impossible for juveniles to meet their energy requirements on a diet of low-energy prey sources.[5] Juveniles one to two years of age were found to have greater energy needs than could be supported on low-quality food sources alone. They couldn't eat enough of the low-quality food to compensate for their high-energy lifestyles. The ingestion of junk food comes at a high price for these fast-swimming, energy-consuming predators.

Seabirds also have been shown to be vulnerable to changes in their food sources from high to low nutritional quality. A diet of walleye pollock, a junk food, is associated with considerable decreases in growth and fitness of juvenile black-legged kittiwakes (*Rissa tridactyla*) and tufted puffins (*Fratercula cirrhata*).[6] Such decreases could have overall negative effects at the population level, since both the quantity and quality of offspring produced by these juveniles will be ultimately decreased by low-quality food. A similar negative outcome of a low-lipid junk-food diet has been demonstrated for juvenile red-legged kittiwakes (*Rissa brevirostris*). A diet of walleye pollock alone resulted in growth retardation, increased secretion of stress hormones, and decreased cognitive abilities in these young birds.[7] The consequences were increased mortality and low reproductive success in those that survived.

A meta-analysis (a broad survey of research on a certain subject) on studies of predators in northern seas suggests that the junk-food hypothesis is an important idea that should factor into ecosystem conservation.[8] Many species are not affected solely by the availability and abundance of their food sources, but by the overall quality of their diet as well. It is unlikely that pollock just tastes good (like the fast-food cheeseburger), but it is quite likely that it is easier to obtain than higher-quality prey.

Climate regime shifts around the world are thought to be at least partly responsible for the restructuring of the food webs in the Gulf of Alaska,[9] the North Sea, and the Baltic Sea leading to a lower availability of high-quality food choices for the top predators inhabiting these areas. For humans, the story is reversed. Our high-emissions, CO_2-producing modes of transport mean that it's easier than ever to obtain high-quality healthy foods with little effort, no matter where you live. Despite these opportunities, most people in the western world still vote for low-quality food, taking the *optimal* right out of the optimal foraging theory.

The Offspring Are Watching!

As parents we aspire to set a good example for our children. From the way we treat other humans and other animals, to the jobs we have, and of course our eating habits, the children are watching. Sometimes I wish they wouldn't watch so carefully. Thanks to my salty snacking tendencies, my three-year-old has developed quite a taste for potato chips. Some degree of teaching and learning with respect to foraging behaviour is critical to all animal species that exhibit parental care.

From birds to lizards and mammals, parents put effort into showing their offspring how to forage effectively (i.e., to obtain nutrition while maintaining a balance with other activities in accordance with optimal foraging theory). In the natural world, juveniles are no longer fed by their parents after weaning or fledging, making the acquisition of foraging skills imperative for survival. If the skills required for food provisioning have not yet been learned by the time juveniles are turfed from the loving arms of mom and dad, there is a very real possibility of death by starvation (and zero possibility of passing on those genetic blueprints).

Young Japanese macaques (*Macaca fuscata*, a widespread member of the primate family) have been found to undergo a concentrated learning phase when it comes to foraging.[10] Both infants (aged 7 to 12 months) and juveniles (aged 1.5 to 2 years) were found to spend blocks of time in intense observation of their elders' feeding behaviour. Being opportunistic feeders, these monkeys have a varied diet, so it is important for youngsters to learn what is edible and what isn't, in terms of both rare food items and food found in unfamiliar habitats. Observations are most often directed towards the mother and can include following the food item from the hand to the mouth, smelling the food in the mother's mouth or hand, collecting dropped scraps of food, and stealing the food directly.

It is important for wild primates to learn foraging skills at an early age, not only because they will be on their own to find food shortly after they're weaned, but also because at this point other adult primates (especially males) will begin to regard juveniles as potential competitors for food, which can lead to violence. Attaining the skills to forage independently, even in unfamiliar territory, is critical for young macaques. It's no wonder they pay such close attention to their mothers.

Appropriate feeding behaviours aren't always learned from the mother. In canaries (*Serinus canaria*), young juveniles most often learn from the father. The fathers feed the young, as the mothers are often busy incubating the next clutch of eggs. Developing efficient seed-handling skills requires simultaneous access to the seed by both the juvenile and the father.[11] He selects and husks the seed and has his child play an active role in the process to facilitate learning. It isn't simply a matter of showing the juvenile what kind of seeds to choose; it's a classic example of learning by doing. The young juveniles then select and husk the seeds in the presence of their fathers, reinforcing the techniques that will be so critical later.

There are a few major differences between the human species and the rest of the animal kingdom with respect to teaching our offspring habits for optimal foraging. First, we often prepare food for our children long after they need us to do so. Despite the fact that most human children do not leave the nest until they're adults themselves (many years later than biologically necessary), many fledgling humans are incapable of efficient foraging and food preparation. Perhaps we should take a lesson from the canary and engage our children in food selection and preparation from an early age in order to prepare them to leave the nest. Second, and this is a big one, humans don't generally forage in an optimal way.

Overconsumption of processed, sugar-filled, high-sodium products is unique to our species. How should we communicate to our infant offspring that instead of the hot dog we ate for lunch, we should have had a salad? The most effective communication in this regard is learning by doing, as with the Japanese macaque and the canary. If you make frequent trips to the drive-thru or have your local pizza place on speed dial, your children are likely to acquire the same kinds of foraging skills. The ease with which suboptimal foraging occurs in our species makes the task of teaching our offspring a good deal more difficult than it is for most other species.

Why Can't We Just Lose Weight?

Members of the human species struggle with body size. Some of us are overburdened with so much extra body mass that tasks such as resource gathering, exercise, and reproduction are more difficult than necessary, while many others of us are only dealing with a few extra pounds that we wish weren't there. Human beings have developed a plethora of dietary strategies by which to decrease body mass in the hope of fitting into the all-important size six. This usually has nothing to do with improving biological fitness.

For most other members of the animal kingdom, a little extra fat on the body is actually a good thing: it acts as insulation in cooler climates and can come in handy as an energy source in times of food shortage. However, too much extra weight can hinder basic activities in animals much the same as it does in our species. There's a reason you've never seen an obese bird flying around—the laws of physics apply universally.

Although *excess* weight gain can hinder flight in birds, there are specific ecological scenarios where packing on a few extra pounds is a necessity. Migratory birds undergo dramatic changes in body composition before embarking on journeys that require several days of flight without stopping for food. The annual migration of the bar-tailed godwit (*Limosa lapponica*) takes it from Northwest Africa to the most northerly part of the Eurasian continent: the Taymyr Peninsula in Russia. This 9,000-km journey is separated into two flight segments with a month-long rest stop at the halfway point.

As you can imagine, substantial body conditioning is required for such a test of endurance.[12] Muscles required for long-distance flight are bulked up (hypertrophied) during the month-long period of residence at the rest stop and reach peak mass as the birds prepare for flight. Tissues involved in digestion are initially hypertrophied as the birds gain weight and then atrophied, since they aren't needed during the rest of the trip. Body fat is dramatically increased during the stopover, both in subcutaneous stores and on an abdominal fat pad. Bar-tailed godwits can intricately manipulate their body tissue makeup in order to embark on the long-distance migration. Equivalent to a person preparing for a marathon or even an Ironman Triathlon, this regimen applies not just to superfit individuals, but to the entire migrating population, male and female, young and old.

Migration is not the only reason a bird might want to manipulate its body weight. Some nonmigratory shore birds go through a period of weight loss by fasting during egg incubation. This phenomenon is described in the programmed anorexia hypothesis.[13,14] Once the eggs have hatched, there are new mouths to feed, and parents often lose weight in order to minimize energy loss during frequent flights to find food for their young. If they're carrying around a lot of extra weight, it will take much more energy for them to fly. Breeding colonies can be far from foraging areas, resulting in a great deal of travel for parents carrying heavy loads of food. Storm petrels (*Oceanodroma leucorhoa*) have been found to decrease body mass via a loss of skin and subcutaneous fat in order to prepare for chick rearing.[15] The decreased body mass of the parents translates to a 14.4 per cent decrease in the energy required for flight, which results in their chicks being given more food. In another species, Brunnich's guillemots (*Uria lomvia*), it has been shown that weight loss can help to improve diving performance, as parents of this species provide their young with resources gathered beneath the water.[16]

It is humbling that birds can alter their body composition according to changes in their ecological requirements while many humans are unable to maintain anything even close to a healthy functioning body weight. In addition to the obvious problems we have in making appropriate dietary choices, this vast dichotomy may be due to the fact that humans are obsessed with weight and size in regard to appearance, rather than gearing our physical size towards functional necessity and maximized health benefits. Instead of maintaining a body weight that allows for efficient performance of our short- and long-term tasks, we rely on outside influences to dictate what our size should be, regardless of whether that size is physiologically appropriate. The absence of such factors elsewhere in the animal kingdom facilitates a *functional* response to weight management as opposed to a *psychological* response, which in my opinion takes the biology right out of the human weight predicament.

Haute Cuisine à la Nature

Most foods that humans eat are altered in some way before we consume them. Our species has developed elaborate ways in which to change the properties of our foodstuffs. Why don't we just forage around like the rest of the animal kingdom? Partly, it's because raw meat or chemically defended plants would make us very sick. *Homo sapiens* are blissfully unaware that the majority of foods available to organisms in a natural setting come with a large cost in terms of energy loss, risk of injury, or poisoning. What's a hungry herbivore going to do in order to survive? Certainly not stop in at the nearest salad bar and grab a plateful of already cleaned, already cut up, readily available veggies. Obligate herbivores face daily challenges from chemically defended foods, and several of them have come up with strategies to cope with food that bites back. After all, when you've been dealt the evolutionary card of vegetarianism, you must come up with ways to make your food work for you. It may not be enough to achieve a five-star rating by human standards, but the act of meal preparation in some species is unmistakable.

Beavers (*Castor canadensis*) increase the palatability of the bark they ingest by leaching it out in the water for several days.[17] There are harmful chemical compounds called phenolics in the bark, so a form of pretreatment is imperative. The beavers create a digestible food product by using the power of water to draw the poisons out. It has been shown that the phenolic content of bark and twigs can be reduced by up to 80 per cent using a leaching method such as this.

North American pikas (*Ochotona princeps*), small members of the rabbit family, have developed alternative ways to deal with the natural toxins in their grass-based diet. They create food piles, called haystacks, to eat from during the winter months when food is scarce. The benefits of storing grasses in haystacks during the winter are many.[18] First, the long storage period gives the toxins in the plants time to break down. By the time the pika consumes the plant, the toxins are at an acceptably low level. Second, these smart little bunnies have been known to add leaves with antibacterial properties to the haystacks, and these are thought to act as a preservative through the winter. In addition to allowing for the breakdown of toxins, they are ensuring the vitality of their food rations by protecting them from invasion by microorganisms.

Like their vegetarian cousins, carnivores must often manipulate their prey in order to remove unpalatable components. In some cases they do this with a finesse that is far beyond just eating around the tough bits. After capturing cuttlefish, dolphins (*Tursiops aduncus*) off the south coast of Australia have been observed to undergo a complex set of behaviours in order to prepare their meal.[19] Cuttlefish (*Sepia apama*) are a cephalopod species closely related to squid; they have an ink sac as well as an internal hardened shell (cuttlebone). Dolphins have been observed preparing them for consumption by releasing the ink, which contains a high concentration of the pigment melanin and other noxious compounds that are hard on the digestive system, and carefully extracting the cuttlebone in one piece, allowing for more efficient ingestion and digestion.

At a rudimentary level, it is clear that a diverse array of animals in the natural world possess the ability to prepare food for ingestion by leaching out poisons (or allowing them to break down), utilizing food preservation techniques, and removing unpalatable components. In the animal kingdom, preparing meals by making foodstuffs more palatable and safer to ingest is directly related to improving biological fitness. In our species, on the other hand, meal preparation can do the opposite, such as taking nutritious potatoes and deep-frying them, making them harmful instead of healthful. Although haute cuisine à la nature may be a few steps removed from the sophisticated meal preparation of the human species, it is meal preparation all the same. By the way, the latest culinary fad to hit the *Homo sapiens* world is both novel and innovative: the Raw Food Diet takes the preparation out of meal preparation and focuses on whole and natural foods. I wonder who came up with such an original concept?

The Crowded Buffet: Wait or Settle?

I'm not a huge fan of the all-you-can-eat buffet. It's the equivalent of a feeding trough for humans. I don't like the crowding; I don't like the line-up of people at the prime-rib station, drooling as their cut of meat is hefted onto their heaped-up plates. I think my behaviour at the buffet is directly correlated to the number of people lingering around a specific area. If I had the place all to myself, I would be more inclined to hit the hot ticket items, but when it's busy and the best parts have been completely picked over, it's probably best to explore the other available options. Optimal foraging theory predicts that when there is intense competition for preferred resources, organisms should increase their diet breadth to include other less optimal items.[20] In this way, biological fitness is maximized by striking a balance between obtaining food and the amount of time and energy required to do so. Prime rib becomes less valuable if there is a twenty-minute wait attached to it. I am in complete agreement with optimal foraging theory on this one. Instead of waiting for a meat slab or fighting over crab legs, I'd rather eat something that may be less 'valuable' but is all mine.

A field full of flowering plants can be thought of as an all-you-can-eat buffet for pollinating organisms. A variety of insects such as butterflies and bees can feast on a plethora of plants, and the plants are only too happy to share their health-giving qualities. After all, when a plant contributes to pollinators, its species will be pollinated, which leads to reproductive success.

Many plants have evolved specialized coloration, form, and scent in order to make themselves more attractive to potential pollinators, not entirely unlike the garnishes, scents, and presentations of the food spread out at our buffets. But what happens when the natural buffet becomes crowded? Do pollinators wait in line for their chance at the hot-ticket flowers or do they follow the tenets of foraging theory and forage for something less exciting?

In an attempt to answer this question, laboratory experiments were conducted to assess the food choices made by the common bumblebee (*Bombus terrestris*) in crowded and uncrowded environments.[21] Artificial plant communities were created with a wide variety of species that included both highly rewarding and poorly rewarding types of food plants. Individually marked bees were followed in two situations, one with only one other conspecific (organism belonging to the same species) present, for low density, or with six conspecifics present, for high density. The number of visits made by the marked bees to each type of plant was recorded in each situation. True to the predictions of optimal foraging theory, the diet breadth of individual bumblebees was increased when the buffet was crowded. Low-rewarding plant species that were visited only 6 per cent of the time in the low-density situation were visited 32 per cent of the time in high density, indicating that not all of the bees were willing to compete for the prime rib. Interestingly, the diet of the bees was most specialized when the buffet was not crowded—in that case only high-quality foods were selected. Although this may be an optimal situation for the individual bee, it doesn't help to maintain diversity in the items available at the buffet. In this context, an increased level of competition may actually work to preserve the biodiversity of the plant community by forcing unpopular plant species to be pollinated along with popular ones.

Many plant-pollinator interactions are opportunistic,[22] meaning that the interactions can vary at different times and places, which has the effect of maintaining the integrity of the system. It may be advantageous for a certain pollinating insect to feed on a particular food type at a specific time, but that insect maintains the ability to select other food sources if necessary. The key is to have the diversity to be able to withstand temporary changes in conditions. When something like mad cow disease causes the popularity of prime rib to take a nose dive, the buffet must be able to compensate by continuing to offer a variety of other foods.

Like the bees, individual humans might be inclined to indulge in a single hot-ticket item if such an opportunity exists, but this strategy isn't optimal for the overall maintenance of the buffet, natural or otherwise. In addition, it isn't optimal for maximizing biological fitness, because organisms should be able to compensate for uncontrollable changes to their food supply. If the ability to do this is lost, the quest to obtain adequate nutrition becomes a lot more difficult. The overall message: a little crowding is beneficial to everyone. There may only be a small spoonful of peas and corn on your overloaded plate, but you'd probably miss them if they were gone.

Come and Get It!

We humans have a variety of ways to remind ourselves that it's time for a meal (as if that growling sound in our stomach weren't enough). There are lunch breaks, dinner bells, and the all-important clock on the wall signalling us to refuel. For those who live in a household where one family member (usually the female) is responsible for meal preparation, the universal *Come and get it!* and *Dinner!* commands are used, loudly, as a signal that the food is ready for consumption. *Homo sapiens* is not unique in its quest to inform others that food is available. Food signalling calls are commonplace in the animal kingdom. A diverse range of organisms, from bees to mole rats, from dolphins to birds, utilize some form of communication with respect to food. How does food signalling help to maximize an individual's biological fitness? The general idea in several animal species is that if you make the call to others when a food source is found, similar generosity should find its way back to you when someone else is the food-finder.

Altruistic organisms are defined as those that live together in groups (not necessarily related to one another) and generally behave in a manner that benefits the entire group (as opposed to behaving selfishly). Working together provides the group members with the ability to detect and exploit a wider set of possible resources than could be located if each individual worked alone. For example, altruistic chickadees forage outside of eye contact with one another, but individuals within the group help each other to obtain adequate nutrition by signalling to conspecifics when a food source is found. Such 'come and get it' signals are deciphered from other types of signalling calls according to their note composition.[23] Observations of the Carolina chickadee (*Poecile carolinensis*) have shown that the first chickadee to arrive at a feeding station produced a call containing a large number of D notes. Once a second chickadee arrived at the station, the number of D notes in the call of the first individual decreased, leading to speculation that the D note serves a recruitment function. In order to test the hypothesis that the D note is critical to the food-sharing process in the Carolina chickadee, researchers played songs containing both a high and a low number of D notes at different feeding stations in order to determine the amount of time that would elapse before birds arrived to feed. The time between the call playback and the first individual's arrival at the feeding station was found to be significantly shorter when the call contained a large number of D notes. The altruistic nature of the chickadees results in a higher level of individual feeding for everyone within earshot of the food recruitment song—a concept foreign to members of other species.

Unlike chickadees, we primates are not reciprocal altruists; most of us are not in the business of providing food to whoever happens to hear us calling. Much to the contrary, the specific signals we use are meant for our close kin only, and there are certain times (such as when there's a new box of chocolates or a batch of freshly baked cookies) that even signals to our close kin may not be loud and clear. Indeed, primates have been shown to exhibit changes in the quality, quantity, and timing of food signalling depending on the context.

In captive chimpanzees (*Pan troglodytes*), individuals were more likely to vocalize about a food source when large quantities of food were discovered.[24] The relative divisibility of the food source was also found to be an important factor contributing to signalling behaviour. When one large piece of food (as opposed to several small pieces) was found, chimps were less likely to signal to nearby conspecifics. If more effort was required in order to divide up the food, individuals were less likely to share.

Tufted capuchin monkeys (*Cebus apella*) take the deceptive use of food-related vocalizations a step farther. Individuals are more likely to signal, using a specialized set of vocalizations and whistles, about a discovered food source when another capuchin is within their visual range. If no one else is around, the individual that discovered the food will take longer to share the information, increasing both alone time at the food source and the overall amount of food ingested.[25] Although more time elapses between finding the food and signalling about it, the capuchins send a signal to their conspecifics eventually, suggesting that there may be negative consequences for acting selfishly. It has been shown in Rhesus macaques (*Macaca mulatta*) that individuals who find a food source but refrain from signalling and are later caught in this act of cheating are treated aggressively by other group members, which ultimately results in less food for the cheater.[26]

Food signalling calls are commonplace in the animal kingdom, but some major differences exist between species. In order to maximize one's chances of survival, it is critical to consider both who is meant to receive the food signal and the probability that the response will be from the targeted parties. If there is a high likelihood that the intended recipients will miss the signal and that unsolicited opportunists will show up instead, then signalling should be kept to a minimum. This generally is not a problem for altruistic species that send signals to all members of the group, but it is a problem for species with more selfish tendencies.

Far from the mere inclusion of D notes in a whistled song, most primates carefully monitor their food-sharing signals. Let's face it: there is no guarantee that altruistic behaviour will ever be reciprocated. In this regard, human nature is clearly reflected in the actions of the chimpanzees, capuchins, and macaques. If you're hungry and only a small portion of food is available, why share? If no one is around and a plate of cupcakes presents itself, what's to stop you from indulging a little before the word gets out? Whether we admit to it or not, we're all likely to take a few bites before sending out the signal that food is available. We may not be as far removed from other members of the primate order as we'd like to think.

Italian, Chinese, or Indian Tonight?

In diverse, multicultural cities, there is a wide variety of choices with respect to food. From sushi to curry, from pizza to chow mein, it's all available in an urban centre. Culinary specialties have arisen in different localities according to a myriad of variables, including food availability, climatic conditions, and level of industrial development. Nowadays, in large cities of the western world, individuals are privileged to have access to all kinds of cultural delicacies through advances in technology such as communications and aviation. As a species, we *Homo sapiens* exhibit a great disparity in the kinds of food we eat, although members of certain cultures are more likely to eat the kinds of foods that originated with that culture. To put it another way, we are a generalist species with respect to food choice, but variation among individuals is high and can actually lead to diet specialization. In this way, we are specializing generalists. It makes sense to be a specializing generalist if you're part of a species with a broad geographic range, and that's certainly true of the globe-trotting *Homo sapiens*.

Herbivorous insects (such as several species of flies, butterflies, and aphids) have been studied extensively in the context of specializing generalist behaviour, since their vegetarian habits require them to have mechanisms for detoxification of chemically defended plants (see Haute Cuisine à la Nature). It wouldn't make sense for an insect to possess the capability to detoxify foodstuffs that it never encounters, since this would represent a waste of energy that could be put to more efficient use. Those that inhabit vastly different parts of the world therefore are physiologically specialized to detoxify the foodstuffs located only in their geographic region. Many herbivorous insects assume a generalist feeding strategy over their entire geographical range but function as specialists with restricted diets at localized scales.[27] In the same way we have pasta in Italy and guacamole in Mexico, many flies and butterflies are generalists and specialists at the same time.

Dusky-footed woodrats (*Neotoma fuscipes*) are small mammals that inhabit certain areas in California. These generalist herbivores live in habitat types dominated by either juniper or mixed coniferous trees. In order to determine the extent to which specializing generalist behaviour occurs in this species, individuals from both habitat types were live-trapped and utilized for dietary choice studies.[28] Woodrats were given a variety of food choices from both the juniper and the coniferous habitats, and it was found that individuals preferred to eat the foodstuffs most common in their home turf (despite the fact that these habitats are not geographically separated or even very far apart). Moreover, they actively avoided unfamiliar chemically-defended plants, possibly because they lack the digestive flora and enzymes required to process them. Like the herbivorous insects, these mammals are generalist feeders that adapt specialist behaviours at smaller scales.

The inability to detoxify chemically-defended foodstuffs provides herbivores with a clear rationale to be specializing generalists, but similar dietary strategies have been observed in carnivorous organisms. The octopus diet is most often assessed through detailed analysis of middens—piles of remains that are really food scraps outside their dens. If examined on a regular basis, the middens provide information about the most recent meals consumed by the individuals living in a particular area.

A survey of octopus (*Octopus vulgaris*) middens in the Caribbean Sea determined that there was a great diversity of prey ingested by the population—seventy-five species of prey. This classifies the species as a dietary generalist, but individuals were found to specialize in certain taxa or even individual species.[29] The specialization was not based on a difference in the ability of individuals to access resources, as it appears to be for the herbivorous insects. All dens were located within close proximity to each other in homogeneous terrain. It seems that for these carnivores, the differences in midden contents reflect individual preferences and handling specializations for specific foodstuffs. Not all humans prefer the same types of food, so why should we expect other animals to do so?

Although the occurrence of the specializing generalist is fairly common in the animal kingdom, a major difference between our species and many others is that we have unparalleled access to the full gamut of foodstuffs enjoyed around the world. Herbivores do not possess the physiological machinery or the proximity to consume the diversity of items utilized by their species as a whole. While detoxification of foodstuffs is not generally an issue for carnivores, they may select foods that are most optimal to them in terms of experience, proximity, and handling time. Such actions lead to a specializing generalist strategy in vegetarians and meat-eaters alike. Humans, on the other hand, can travel to a faraway land to experience another culture and its diet. We can sample a wide variety of cultural specialties prepared locally with no ill effects on our biological fitness. Unlike most other organisms, we have the luxury of universal choice. We can select either a generalist or a specialist strategy, and that makes me quite happy to be human.

Controlling the Crops

Humans have perfected the art of farming. We have learned to create beautiful, healthy crops of fruits, vegetables, and meats to feed our masses. I live in a farming community, with vast fields of dairy cattle and the grasses and corn to feed them. One is always aware of fertilization time, when thick black manure is spouted from sprinklers for hundreds of metres. Most of the landscape in the area where I live has been altered so as to maximize output from the plants and animals dwelling there. Once an area has been cleared for farming, care is taken to ensure that plants and animals other than those of the desired crop are suppressed.

Although the benefits of farming are felt far and wide among the human population in order to optimize our diet with little effort, the functional result of farming on the natural landscape is to suppress the natural biodiversity of communities that would otherwise inhabit such areas.

As in this farming scenario, a mutualistic relationship exists among different species of ants and aphids. Aphids are small herbivorous insects that spend their entire life cycles on specific host plants. They excrete a sugar-rich substance called honeydew that is a prized resource for various ant species. In exchange for this food source, ants provide the aphids protection from predators[30] in the same way humans give dairy cattle a safe environment for feeding in exchange for the milk they provide. It has been documented that, in some particularly aggressive ant species, aphids are herded onto specific types of plants that are more productive and have positive effects on honeydew production.[31] Ant tending has been shown to increase honeydew production by up to 50 per cent,[32] indicating that ants stand to realize large benefits by ensuring that their own aphids have a safe and nutritious habitat.

The aphid *Chaitophorus populicola* and the ant *Formica propinqua* exist in this mutualistic relationship: the aphids are protected by living in close proximity to the ants, while the ants gain the nutritional benefits from the honeydew. Aphid numbers drop by 88 per cent when their tree habitats are located more than six metres from an ant mound, and no aphid colonies at all are found further than ten metres from an ant mound.[33] In other words, without protection from their tending ants, the aphid populations are much more likely to be wiped out by predators. The aggressive ants actively remove potential predators from aphid-inhabited trees, including other herbivores and competing ant species. However, in providing the aphids with this safe habitat, the ants affect the makeup of the communities present on the plant.

Overall species richness and abundance is greater on trees without aphid-ant mutualists than on trees where they are present. When aphids were experimentally removed from trees and the aggressive ants therefore abandoned them, there were increases in overall species richness (31 per cent) and arthropod abundance (26 per cent), demonstrating that ants were effectively clearing the trees of both predators and competitors (thereby decreasing biodiversity) to ensure the success of their honeydew crop.[34]

From an optimal foraging point of view, farming makes sense. Instead of all individuals hunting and gathering as independent units, taking steps to ensure that a viable food source is reliably available to all members of a population in a specific area seems like a good idea. Tending ants have a pretty good thing going when it comes to their honeydew-producing livestock, despite the negative effects on other members of the community. However, it's important to note that the ill effects on biodiversity are at the level of the individual tree.

At a landscape scale, the integrity of the natural system remains intact, and the organisms shunted from the ant-tended trees take up residence nearby. Ants haven't turned honeydew farming into a global commodity, as humans have with our farm products. Farming provides ample nutrition to most humans, but the large-scale land clearing required for such practices, especially in diverse areas like tropical rainforests, reduces biodiversity until the loss is catastrophic for all other species involved.

"WHEN HIS BEADY EYES LOOKED INTO MINE, I KNEW THE GAME WAS AFOOT!"

Waste Not, Want Not

I have a strange set of neighbours: a house full of middle-aged bachelors. Every year they go on a hunting trip and come home with a large deer or moose. They set themselves up in the garage and butcher it, much to the chagrin of neighbours walking by. Last summer they came to our annual block party with deer steaks for everyone to try. They weren't bad.

When I tell people about my neighbours' practices, I'm usually met with a disapproving attitude, but honestly I kind of admire them. I appreciate that they are utilizing the entirety of the animal they kill and that they go through the process of butchering it themselves. I think that if all humans had to butcher their meat before eating it, there would be a lot more vegetarians among us. This brings me to another point about humanity that I find thoroughly upsetting: trophy hunting. If you have no plans to utilize what you are going to kill to improve your biological fitness, then in my opinion there is no reason to kill it. Other animals are generally a great deal more efficient and purposeful with respect to maximizing biological fitness than the confused *Homo sapiens*, so I was surprised to learn that wasteful killing is not restricted to our species.

Surplus killing is common in the insect world, but the reasons behind it are unclear. Wasteful slaughter of mosquito larvae by predatory midges (*Corethrella appendiculata*) is a frequent occurrence in the tree-hole habitats in Southern Florida where these two organisms cohabit. Laboratory experiments were conducted to determine exactly when and how the killing occurs.[35] Individual midges were reared in the laboratory, and once a certain developmental stage was reached, they were placed in experimental containers with a specific number of mosquito larvae. The number of victims that were killed and eaten in episodes of traditional predation and the number that were killed and left aside were noted in all cases. It turns out that only the oldest midges (the fourth instar that occurs directly prior to pupation) were guilty of the wasteful murders. The attack sites on the victims were vastly different depending on the intended outcome. Individuals killed for ingestion were most often attacked on the larval head, which is most efficient if the prey is to be eaten whole, as opposed to the surplus slaughtered individuals that were attacked on the thorax or midsection. The different killing technique suggests that this behaviour is *not* a case of predation gone wrong. It seems to have evolved independently, leaving scientists puzzled.

Several hypotheses have been put forward as possibilities for the evolution of such behaviour. The *vulnerable pupal hypothesis*[36] suggests that an organism that is about to pupate (as the guilty midges were) and enter into a state of vulnerability might attempt to destroy potential predators preemptively. Unfortunately, the earlier study did not support this hypothesis, as the murdered mosquito larvae were young juveniles and unlikely to become predatory as the midges entered pupation. Another hypothesis suggests that wasteful killers might be attempting to decrease potential competition for other prey.[37] The tree-hole habitats where the larval midges and mosquitoes live are closed, meaning that the insects, who are entirely aquatic during these life stages, are not free to move elsewhere. Living in such close quarters with one's closest competitors provides the rationale for the 'decreasing your competitors' hypothesis, but that was not supported by the midge/mosquito study either. Pupation is a nonfeeding developmental stage. Decreasing the population of one's closest food competitors when one is about to enter into a nonfeeding state would be a waste of time and energy.

To date, then, the biological reasons behind nonconsumptive prey killing in the insect world are not clear, but the tree-hole CSI teams continue to investigate. What we do know is that the discarded carcasses left in tree-holes or elsewhere are ultimately consumed by a multitude of other organisms eagerly awaiting their chance at a free meal. In this sense, while the killing may be a surplus or wasteful event for one organism, there are plenty of others that will utilize the corpse for nutrition or habitat.

This makes me ponder the reason that surplus killing occurs in our species. In my humble opinion, it is often the testosterone-driven need to hold power over other organisms. To add insult to injury, the taxidermized wastes of countless creatures that lost their lives to trophy hunters did not get the opportunity to be recycled through the food chain. Their deaths were in vain, and our species is the ultimate food waster.

Chapter 2

Staying Healthy in a Cruel World

Got Worms?

Parasitic infections are an unfortunate part of life for most animals. We are all susceptible to a wide range of infectious invaders from single-celled protozoa to lengthy tapeworms. For *Homo sapiens,* parasitic infections are generally preventable by means of hygienic conditions and well-cooked food. However, there are always exceptions to the practice of vigilance. Accidental exposure can occur through individuals whose standards of hygiene or food preparation differ from our own or through direct contact with parasites during travel to a foreign country. It is important for humans to remain attentive to the possibility of parasitic infection.

Such infections are remarkably more common in our primate cousins, quite likely due to the fact that exposure is much more likely when one exists without the (unnaturally sterile) human standards of acceptable hygiene. How do the great apes cope with nematode or tapeworm infections? When one's own physiology is not enough to rid the body of an intruder, behavioural interventions must take place, so much so that the impact of parasitism on the host is considered to play an important role in the evolution of behaviour in the animal kingdom.

When plagued with a parasitic infection, most commonly the nematode *Oesophagostomum stephanostomum* or the tapeworm *Bertiella studeri*, chimpanzees from several independent populations on the African continent have been observed to eat whole leaves from various plant species. The leaves are carefully folded and then swallowed slowly without chewing. They remain largely undigested on their journey through the gastrointestinal tract, and extruded leaves carry with them many of the offending parasites.[38] A common characteristic of all leaf species utilized in this way is the presence of trichomes, or rough patches on their surfaces. The trichomes, along with the nooks and crannies created from folding the leaves, serve to trap the adult worms from the gut, helping to clear up the infection.

Heavily infected individuals often consume the leaves early in the day, upon waking. This makes sense from a healing perspective: the physical irritation produced by the bristly leaves on an empty stomach serves to stimulate the digestive tract to remove potentially damaging substances.[39] In other words, eating irritating leaves on an empty stomach gives you diarrhea. This extra fecal material is often found in association with the undigested leaves, which contain not only adult parasites, but larval cysts as well. Infected individuals have been observed to ingest from 5 to 55 leaves, often repeating the process several times, leading to the conclusion that this technique is effective in its treatment of parasite infection. Indeed, this kind of behaviour has been documented in bonobos and lowland gorillas, as well as in other vertebrates such as brown bears and snow geese. The practice is ubiquitous and effective at clearing up infections across the animal kingdom.

Humans, on the other hand, have devised other ways in which to treat parasitic infection, thank goodness. We can do this prophylactically through vaccinations or therapeutically through the use of medications. The general technique for treatment of maladies such as roundworm, hookworm, pinworm, and tapeworm in *Homo sapiens* is one of differential toxicity. The therapeutic drugs are toxic to our own bodies, but they are more toxic to the bodies of the parasites and hence effective in removing them from our system. Side effects from such drugs are common due to their general toxicity but not life-threatening in most cases, and we don't have to suffer the discomfort of swallowing large numbers of neatly folded bristly leaves for breakfast. The chimpanzees and other apes that utilize this technique must have evolved some serious antigag reflexes in order to ingest the leaves without chewing them. I suppose when no other alternative is available, the short-term discomfort is well worth the survival benefits to be gained by removing the parasitic invaders. On this one, we humans are lucky to have developed more palatable alternatives.

Slather It On!

A dab of this, a drop of that—we humans are consistently covering our bodies with creams, oils, and ointments of all kinds. From moisturizer to perfume, from antibiotic ointment to soothing muscle rubs, we've got our bases covered when it comes to finding a perfect balm for what ails us. This is an extremely good thing, considering that our outer covering provides us with precious little protection. Without the tough integument of an elephant or amphibian, or the protective coat of many other mammals and primates, we're at risk for a plethora of irritations or microorganismal attacks on our delicate skin. Although most organisms are not as easily exploitable as *Homo sapiens* when it comes to assaults on our outer surface, such attacks occur commonly in the rest of the animal kingdom.

Self-anointing is defined as any behaviour in which active compounds from one organism are purposefully transferred to the skin or fur of another. Self-anointing is widespread in the animal world and is thought to serve several functions, including repelling insects and ectoparasites, treating wounds and infections, and combating inflammation.[40] For example, capuchin monkeys (*Cebus* species) in Venezuela have been observed to self-anoint with secretions from millipedes (*Orthoporus* species). The chief defensive compounds (poisons) found in the millipedes are benzoquinones, which function to deter their predators. The use of millipede-derived benzoquinones by the monkeys is hypothesized to minimize the bite-frequency of mosquitoes by reducing the frequency of their landing and limiting the time they spend on a particular feeding target.[41] This behaviour is important to prevent the spread of the yellow fever and dengue viruses carried by the pests, and it therefore helps the capuchins to survive.

Another use of insects for their defensive compounds is *anting*, or self-anointing using carpenter ants, whose defensive secretions of formic acid can be highly potent against predators and pests. Anting has been observed in a number of bird and primate species and is thought to function primarily as a defence against ectoparasitic infection. Capuchin monkeys (*Cebus apella*) in Brazil have been observed to engage in active anting behavior.[42] They sit near nests of carpenter ants and pick unlucky individuals to rub onto their fur, or they allow the ants to crawl on their bodies and then rub them in as if they were little packets of body lotion. Capuchins primarily undertake anting behaviour at the times of the year when tick populations are prevalent. The concurrence of anting behaviour with the abundance of ticks at certain times of year and the fact that self-anointing does not include eating the ants both point to the conclusion that the formic acid is a useful deterrent against tick infection.

In addition to self-anointing to deter ectoparasites and pathogens, several other therapeutic uses have been described. Orangutans (*Pongo pygmaeus*) from the swamp forests of Borneo have been observed to self-anoint in order to treat inflamed muscles. Leaves of plants in the genus *Commelina* are used by many indigenous tribes for muscle pain, sore bones, and swellings, and they appear to serve the same purpose in the orangutans (one wonders who thought of it first). Over 10,000 hours of behavioural observation of orangutans in Borneo revealed that adults pick several leaves from the top of the plant, chew them for three to five minutes (presumably to release the saponins in the leaves), and apply the lather-like mixture in a careful, purposeful way.[43] These self-anointing rituals could take more than 30 minutes and involved concentrated application around specific muscles and joints. No species of *Commelina* plants are food sources for orangutans, indicating that the plant is used only for its medicinal qualities.

It would be simplistic and presumptuous of humans to assume that we are the only organisms to utilize the medicinal qualities of compounds to treat topical infections and other ailments. In fact, researchers have long recognized that there is a clear link between self-medication in animals and ethnomedicine, the traditional medicine practiced by local and indigenous peoples. Much of what we know about the use of natural compounds as medicine comes from observing our animal cousins in various situations. *Homo sapiens* in the western world have taken the use of such compounds to the next level: we synthesize pure and concentrated forms that are much more effective and specialized in function. Most of these help to protect our weak integument from a host of insects, parasites, and other external assailants, but we are quite likely to overmedicate ourselves to the point that without such intervention our poor bodies cannot deal with the intruders. Natural? No. Human? Yes.

Aromatherapy

The aromatherapy business is booming in the western world. Essential oil products that claim to rid us of assorted ailments are available in a wide range of forms. Soaps, sprays, and candles containing aromatic compounds from various plant species promise to alter our mood and improve our health. The big question is this: Do they actually do what they are supposed to do?

One can argue the merits of aromatherapy for humans both pro and con, but there is no question that its use in nature is widespread and effective. The natural aromatic compounds utilized in the animal kingdom don't come in bottles, or with promises to relax and invigorate, but their positive effects on survival are demonstrably clear.

Nesting birds are a prime example of organisms that capitalize on the benefits of aromatherapy. A bird's nest is a welcome environment for pests and parasites, and helpless nestlings left alone while parents are out foraging are an easy target. This combination of factors represents a dire scenario for the newly hatched birds, but several species have come up with solutions to protect their nests and nestlings from invaders. The nest protection hypothesis[44] describes the behaviour of adult birds who adorn their nests with aromatic greenery. The secondary compounds in some plant species may serve to repel parasites or to mask the chemical cues that parasites use to find their hosts. For example, many nesting birds add sprigs of yarrow (*Achillea millefolium*) to their nests. The plant is rich in volatile compounds and is thought to aid in the reduction of parasite loads on young chicks. Indeed, when nests of the tree swallow (*Tachycineta bicolor*) were experimentally supplemented with yarrow, the flea load in nests was reduced by nearly 50 per cent.[45] The addition of yarrow to the nest represents an increased biological fitness to the parents through the protection of their offspring. Without such supplementation, the efforts to pass on their genes to the next generation could be considerably compromised.

In another example, adult Corsican blue tits (*Parus caerulens*) have been observed to bring a combination of aromatic herbs to their nests at night, suggesting a possible protection against a nocturnal pest. The ornithophilic mosquito *Culex pipiens* is common in Corsica and can be a vector of avian malaria, making it a substantial threat to the survival of blue tits in the area. Biologists tested the effectiveness of the five aromatic herb species utilized by the blue tits in a laboratory experiment[46] and found that the combination of herbs was a much more effective deterrent than any of the herbs alone. The aromatic potpourri utilized by the blue tits at night was shown to be an effective deterrent against the virus-carrying mosquitoes.

Birds aren't the only organisms to utilize the powers of aromatherapy in order to keep pests away. Wood ants (*Formica paralugubris*) have been observed to incorporate pieces of solidified conifer resin into their nests. The active aromatic compound in the resin (turpentine) is found by the ants through its scent and is utilized preferentially over other kinds of structural materials in the ants' nest, such as sand grains and wood.[47] In addition to aromatic properties that serve to protect the nest from parasitic microorganisms, the resin contains several antimicrobial compounds that deter invading bacteria and fungi.

The case for aromatherapy in nature is strong and convincing. Why don't humans follow in the footsteps of the invertebrates and birds that capitalize on the benefits of natural aromatics? It may be that we are too far removed from the need for such compounds to contribute directly to our survival. Besides, would the average human even know which plants to use or where to find them? Would we use them in the correct amounts, or would we make ourselves sick with the scent of too many overpowering fragrances? It may be best to leave the birds and ants to do what comes naturally to them while we use our own watered-down version of aromatherapy, which may or may not contribute to our biological fitness. My mango-coconut air freshener may attract the very pests that natural aromatherapy aspires to dispel, but luckily I can obtain creature-repelling poisonous bug spray to fend off any pests that dare to come too close. Ahhh, the complexities of human nature!

Flu Season

Humans inhabiting temperate climates are subjected to a wide variety of environmental conditions each year. Although the fall is a beautiful season, it is a precursor to the unforgiving winter months. Days get shorter, temperatures get cooler, and the amount of precipitation increases. The cold temperatures and wet days contribute directly and indirectly to an increased susceptibility to the cold and flu viruses that rear their ugly heads at this time of year. Although our species manages to shield itself effectively from the elements by creating warm indoor spaces (a luxury not available most animals that inhabit the temperate climate with us), we increase our chances of pathogen transmission by being in close contact with other *Homo sapiens* in such areas. High densities of any animal in a confined space equals high transmission risk of viruses working their way through the population.

The symptoms of cold and flu season are accentuated by the extra energy we must expend to keep warm outside and by the increased exposure to pathogens that we face by remaining inside. Lucky for us, we have developed a number of methods to stay healthy and functioning despite the increased odds of viral exposure during flu season. Each year many of us inoculate against certain viral strains by getting a flu shot. While the shot doesn't guarantee perfect health, it provides us with additional ammunition to fight off a selection of nasty bugs that work their way through the population every winter.

Other animals also experience conditions that merit differential use of the immune system during the year, but they aren't lucky enough to have a local clinic providing them with a flu shot in order to battle the increased demands on their physiology. Long-distance migratory birds experience a wide variety of environmental and pathogenic conditions all year. For example, red knot shorebirds (*Calidris canutus*) exhibit two general strategies through the year with respect to immune function. Scientists have identified these strategies by examining indicators like the microbial killing capacity of the blood, leukocyte concentrations, and levels of natural antibodies.[48]

The first strategy is a state of high immune function that is exhibited during times of the year when susceptibility to pathogenic infection is high. During migration, the red knots fly through vastly different geographical areas, each with its own set of unfamiliar pathogens, creating a need for a heightened state of immunity. In order to complete migration successfully, it is imperative to be able to fend off the assorted pathogens one meets along the way. In addition, during stopovers en route, the birds feed in dense groups, increasing the risk of disease transmission between individuals, much like the risk we experience in a crowded indoor venue during flu season. The second general immune strategy in red knots is a low-cost one favoured when the immune system is less likely to be threatened or when energy demands for other life processes are higher. Once migration is complete and breeding season begins, the red knots establish large territories. Not only is there a smaller risk of pathogen exposure, there is also a greatly reduced chance of swapping pathogens between individuals when the others are far away in their own territories. Demands on the immune system are relaxed in order to conserve energy for reproduction, moulting, and other vital processes.

Unlike *Homo sapiens*, who depend on external sources of immunity such as flu shots, other organisms in the animal kingdom must balance their energy budgets and immune systems without external help. The physiological functioning of the red knots' immune system changes according to their ecological requirements at different times of the year. The varying lifestyle demands and changing environmental conditions they experience over the course of a year have facilitated the evolution of awe-inspiring fine tuning. Despite the fact that some members of the human species have the brain power to develop inoculations and antibiotics, it seems to me that these innovations allow our bodies to become less and less able to cope with naturally occurring infections. Is this a good thing?

Feed a Cold…

Sometimes it seems as though the smallest medical ailments can be among the most annoying. Having a mild bacterial or viral infection can make even the mightiest *Homo sapiens* fall like a baby into his mother's arms. Everyone feels miserable when they've got a cold. We're stuffy and snotty, our throats ache, and our energy level is close to zero. Although we haven't got much of an appetite, we're constantly reminded by our mother's voice in our heads: *Feed a cold!* I try to repeat this mantra when I've been struck down by a cold, to fill up on chicken soup and other foods that might return my body to some semblance of its former self. It makes sense that one should alter one's dietary habits when inundated with an infection. Physiological immune responses come at a high cost in energy and protein requirements, and the body needs extra fuel to make up for resources used up in fighting the infection. Replacement needs to come from some external source.

Bacterial and viral infections are ubiquitous in the animal kingdom; we humans aren't the only animals to bear the wrath of the annoying cold or flu. In fact, it turns out that other species know all about the 'feed a cold' rule, despite the fact that their mothers didn't alert them to it. African armyworm caterpillars (*Spodoptera exempta*) can adjust their dietary composition according to the requirements of their immune systems.[49] The diet of these organisms generally consists of two major components: protein and carbohydrate. The latter is preferred and more highly represented. In order to assess how the caterpillars' dietary preferences would shift in the face of a bacterial infection, larvae were injected with a sublethal dose of *Bacillus subtilis*. It was hypothesized that individuals with bacterial infections would forgo their preferred carbohydrate diet in favour of a higher ratio of protein in order to boost immune function. This is exactly what was found.

Bacterially challenged larvae selected a diet rich in protein, whereas uninfected larvae did not. In subsequent experiments, it was found that the survival rate of bacterially challenged larvae that ingested higher levels of protein was substantially higher than those that were prevented from doing so. These results confirm that there is a substantial protein-based cost to fighting and surviving a bacterial infection. Replenishing lost protein reserves in the face of illness is imperative; in fact, caterpillars on the high-protein diet exhibited increases in antibacterial activity and immune function.

A similar result was obtained from an experimental study in which caterpillars of another species (*Spodoptera littoralis*) were experimentally afflicted with the Nucleopolyhedrovirus.[50] When individuals were allowed to self-select their diets, those plagued with viral infections increased their relative intake of protein compared to control (healthy) individuals. The increased protein demands on the body during an infection are thought to be responsible for this compensatory shift in diet.

The caterpillars from both studies were able to improve their chances at survival through a dietary shift designed to increase protein when inundated with an infection. Further, the caterpillars utilized in the bacterial infection experiments were raised in isolation from very early in their development, making it impossible for them to have learned about the role of dietary protein in boosting immune function by association with others or from their moms. These tiny organisms managed to select a diet that significantly improved their chances at survival without any kind of outside intervention.

Would a *Homo sapiens* be able to make the same choice? Would we really eat more of that protein-rich chicken soup if we hadn't been specifically told to by our mothers or grandmothers? We don't often associate a high level of intelligence or sophistication with animals without backbones, although we know they display a high level of coordination and efficiency when it comes to everyday function. The innate ability of the caterpillars to select a diet that boosts immune function when they are plagued with a bacterial or viral infection (despite the fact that other, more preferred foodstuffs are still available) is a characteristic that our species seems not to possess. Caterpillars 1: Humans 0.

Oxidatively Stressed?

Antioxidants are the new pink. It seems that every week there's some kind of new and improved antioxidant-rich superfood out there that we should be eating: pomegranates, blueberries, and blackberries, to name a few. Antioxidants serve an important purpose in the bodies of the animals that consume them. These plant-synthesized compounds function as scavengers of highly reactive oxygen compounds (ROEs). Commonly referred to as free radicals, ROEs are constantly being produced as by-products of stress, healing, and other energy-consuming processes. They can cause substantial damage to DNA, proteins, and lipid molecules, which invariably leads to premature aging or other degenerative diseases. The bottom line is that ROEs have negative effects on survival and fitness, and it is therefore important to counteract their activities.

One of the best methods to do this is by consuming dietary antioxidants. Anthocyanins and carotenoids are two that effectively combat oxidative stress, and they are found in high levels in ripe fruits. Although all fruits contain some antioxidants, some pack a lot more of the coveted molecules. The good news for the visually sensitive *Homo sapiens* is that the antioxidant value of a certain type of fruit often can be gauged by its color.

Nutrition experts advise us to keep our plates filled with assorted colors in order to maximize the nutritional benefits of our fruits and veggies and also to include plenty of dark, rich colors. There is good science behind these recommendations: the dark-coloured fruits and veggies contain a lot of antioxidants. Dark blues and purples are among the best colors for antioxidant value and are the colors of many of those superfoods. How clever is the *Homo sapiens* to have discovered the correlation between a food's color and its nutritional content! Do any other animals select a diet based on color for that reason? Is it possible that they could be so astute as to utilize color to incorporate antioxidants into their diets without the informed opinion of a college-educated nutritionist?

European blackcaps (*Sylvia atricapilla*) are seed-dispersing (i.e., fruit-eating) birds found across Europe. The overall diet of the blackcap contains a wide variety of fruits from more than 60 tree species, but the specific food choices made by these birds depend on a number of factors. Anthocyanins are one of the two most potent types of ROE scavengers (antioxidants) found in the fruits consumed by the blackcap. In an analysis of the fruits that make up the diet of this frugivore, biologists found a statistically significant relationship between the reflectance spectra—the colors—of the fruits and their anthocyanin content.[51] In other words, there are differences in the antioxidant content of various fruits in the diet of the blackcaps just as there are in the human diet, and these differences are reflected in their color. Researchers then used the avian eye model, derived from the spectral sensitivities of birds' eyes, to determine whether the birds can discriminate among fruits of different colors (and antioxidant contents). Once it was determined that blackcaps can distinguish among fruits on the basis of color, experiments were designed to test whether they actually select the superfoods over their less nutritious counterparts. When given a choice between antioxidant-rich (colorful) foods and antioxidant-poor (drab) foods, the birds overwhelmingly selected the former. European blackcaps actively increase their intake of dietary antioxidants when foods containing them are available. Go blackcaps!

It seems simple, doesn't it? Antioxidants combat oxidative stress, so we humans should be all over antioxidant-rich foods when they're available, and in the western world, they're always available. Not only do we know that foods rich in color contain amore antioxidants (how the blackcaps determined this, I'm not entirely sure), we also know that the spectral sensitivities of our own eyes allow us to differentiate among foods of varying colors, except for those of us who are colorblind. Like our avian cousins, do we use our physiological and cognitive abilities to select the most antioxidant-rich foods available?

I think you know where I'm heading with this. Despite being told what foods are healthy and should be consumed in order to maximize our health and fitness, *Homo sapiens* quite often do just the opposite (see Junk Food). Are our brains too complex to understand the spectacular simplicity of it? Some foods are healthy and will improve your state of being if you eat them. You can find these foods by looking for brightly coloured fruits and vegetables in the supermarket or in your garden. Did you ever notice that there are no brightly coloured foods in your takeout lunch from the local fast-food joint?

Don't Kill the Fever!

One of the most common ways that endothermic animals such as *Homo sapiens* deal with fungal and bacterial invaders is by running a fever. Elevated temperatures in our tissues represent the body's attempt to kill off offending organisms, even though a fever is not without its cost to our emotional well-being. Any parent who has sat at the bedside of an infant child overnight during a battle against a high fever knows this physiological reaction can be as scary as the invaders.

The fever is essentially our body's way of fighting back against the infections that attack it, so in many cases a little fever is actually a good thing. It shows that the body is responding appropriately to a foreign invader, who will soon fall victim to the uncomfortable or lethal temperatures to which it's exposed. The ability to produce a fever in response to infection is common in endotherms, and this process is an advantageous product of evolution. But what about organisms without this built-in physiological mechanism? Ectothermic animals such as reptiles and insects cannot regulate their own body temperatures and consequently rely on their environment to help them out. One might assume that such organisms cannot utilize the health benefits gained from having a fever, but one would be wrong.

Behavioural fever defines the actions of ectoderms that seek to elevate their body temperatures beyond that of healthy individuals when they fall victim to an infection. This can be achieved by basking in areas of elevated temperature, orienting towards solar radiation, shifting from a nocturnal to a diurnal lifestyle, or increasing activity levels, such as extraordinary flight by insects.[52] Behavioural fever has been reported in response to pathogenic infection for a diverse number of organisms, including scorpions, crustaceans, amphibians and reptiles, suggesting that it is widespread for therapeutic use.

Its effectiveness has been confirmed by research examining the process in migratory locusts (*Locusta migratoria*). Individuals were inoculated with a fungal infection or with a control of distilled water and allowed to thermoregulate (hang out) in an experimental arena with regions of varying temperatures.[53] The locusts could move freely between three temperature zones, and they were observed several times a day for up to eight days after being infected. Following the behavioural observations, haemolymph was collected in order to determine the progression of the pathogen and to relate it to the temperatures in which the locusts spent the most time. Those exposed to the pathogen before the experiment exhibited behavioural fevers more frequently than those that had not, and exposure of only four hours a day to the higher temperature had a dramatic positive effect on their survival (85 per cent). The infected individuals did not remain indefinitely in the high temperature areas; they exposed themselves to the high temperatures in frequent periodic bouts, suggesting that this was sufficient to keep the infection at an acceptable level. Although it did not rid the body of the fungus entirely, the therapeutic effects of the behavioural fever allowed infected organisms to continue feeding and otherwise behave normally. It also increased their rate of survival. The capability of ectotherms to control their fevers to such a fine-tuned extent represents a significant advantage in coping with pathogenic infection and maximizing survival. By specifically controlling their exposure to high temperatures, infection could be managed without concern for the damaging effects of fevers gone awry.

There are times when the high fevers produced by endothermic organisms such as humans are too high for our own good. They kill off infection, but fevers that are too high present a risk of physiological damage to the host. For this reason *Homo sapiens* has developed ways to cool off a fever, medication being the fastest and most effective. Unfortunately, members of our species often diminish a fever before it reaches a high enough level to do its job, bringing to a halt its efforts to get rid of the infectious organisms in our bodies. Many of us pill-popping westerners need to remember that a fever is a natural defence mechanism provided by our physiology. The fever is not the enemy! It is necessary to prevent temperatures from going too high, but we need to leave the medication on the shelf until it's really needed and allow the body to do what comes naturally.

Keep It Clean!

Injuries happen. Whether you're a clumsy human or a lucky animal who has narrowly escaped the clutches of a predator, surface abrasions and wounds are common. When surface maladies occur, it is vitally important for all animals to treat them in order to avoid infection. For humans in the western world, this is fairly easy to do. The use of sterile equipment, antibiotics, antiparasitic and antifungal medications, and the ability to care for injuries without the threat of predation gives us ample opportunity to heal. Such luxuries aren't available to the other 98 per cent or so of organisms with whom we share this planet. Animals must rely on natural remedies and their own physiological power in order to heal themselves from wounds and to avoid invasion by the opportunists who are actively seeking a place to fester.

Tropical fish living in coral reef environments have an effective strategy when it comes to caring for wounds and dealing with ectoparasitic infections: the cleaning station. Some types of fish and invertebrates perform a cleaning procedure on other kinds of fish, not entirely unlike the treatment we receive at a medical clinic to clean and care for a wound. Client fish (the ones to be cleaned) assume a specific posture at a cleaning station (an area of a coral reef known to be frequented by cleaning fish) in order to solicit a service.

Clients have been documented to do this up to 144 times a day,[54] making it clear that such interactions are common in the coral reef environment. This setup is viewed as mutualistic, with the clients benefitting from the cleaning and the cleaners benefitting from a free meal. Blue Tangs (*Acanthurus coeruleus*) in the Caribbean experience a high frequency of minor surface abrasions from brushes with sea urchins, hard corals, and larger predators. However, such injuries have a low probability of becoming infected and resulting in further detriment to the health of the fish. How do they heal so effectively without becoming infected? Cleaner fish bite at the periphery of the wounds and at dangling muscle fibres, effectively cleaning off any necrotic or infected tissue (one fish's rotting tissue is another one's meal). Healing of injuries is rapid and complete, with a high rate of recovery for all wounded fish, including those with severe abrasions. Individuals with significant surface damage have been documented to spend much more time at cleaner stations when their wounds were fresh (25.4 minutes per hour) than when they had healed over (1.6 minutes per hour).[55]

Cleaning stations also provide a medical reprieve for fish that are heavily infected with ectoparasites, a common occurrence in the aquatic world. The attending cleaners selectively prey on the ectoparasites plaguing their clients, as opposed to simply grazing on their surfaces at random. When large surgeonfish (*Ctenochaetus striatus*) in Australia's Great Barrier Reef were experimentally traumatized with an ectoparasitic isopod on one side of their bodies so that they sustained high rates of infection on that side and no infection on the other, cleaners spent more time working on the infected sides of the fish.[56] The active removal of parasites and the time spent on heavily infected areas indicate that this kind of medical treatment has an intended purpose for both the client and the cleaner. In another experimental manipulation, the cleaning process was disrupted by holding client wrasse (*Hemigymnus melapterus*) in large cages without access to their attendants. Individual wrasse in the cages carried a four-fold higher parasite load than those that had access to cleaning services,[57] indicating a strong disadvantage to the clients without access to medical intervention.

These examples demonstrate that the active removal of parasites is an extremely important component of cleaning symbioses, moving far beyond the simplicity of animals picking food off of one another. Reef-dwelling fish have developed effective ways in which to treat surface wounds and infections. The solution: keep it clean!

Unfortunately, the human version of 'keep it clean' is a little more drastic. To the great majority of *Homo sapiens*, keeping it clean equates to killing all the other organisms in the vicinity of the offender, even if it means harming ourselves in the process. We don't realize that when we kill off all of the bad bacteria and other microorganisms, we also kill off the good ones that keep our bodies functioning as they should. The delicate balance that keeps our machinery working is constantly being disturbed and reset by our overuse of antibiotics and our need for sterility in our microhabitats—our homes. Moderation is not popular with most of western society, and the tendency to overkill carries dangerous consequences for our bodies' natural ability to cope with common and uncommon ailments.

Sex or Sick?

The phrase *Not tonight, Honey, I have a headache* is thought of as an excuse given to a partner in order to avoid engaging in sexual behaviour. However, there are times when the phrase is actually true. If one is feeling under the weather, getting busy beneath the sheets is a low priority (biologically or otherwise). Not only are we at an attractiveness disadvantage when we're sick (sex appeal is generally masked by symptoms like mucus and vomiting), we're at risk of compromising our health even more if we engage in sexual activities that take energy away from helping our ailing bodies to heal. For humans, this isn't a problem, since we purposely miss most chances to propagate our genetic lineage anyway. We stay in bed for a few days or a few weeks and utilize all available resources to gather strength and nurse the ailing body back to health. However, for most others in the animal kingdom, a compromise in health is a much bigger problem. If, for example, an organism becomes ill at a time of seasonal reproductive activity, that animal has a difficult choice to make: sex or sick? To lose a shot at reproduction in the animal kingdom can be equated to being functionally dead, so this isn't a decision to be made lightly.

Tree lizards (*Urosaurus ornatus*) are an aggressive species, and in the wild they are often found with wounds on their external surfaces. It is therefore not uncommon for these organisms to make decisions about survival and reproduction in a wounded condition. In experiments examining this conundrum, scientists found that the availability of resources played an important role in the 'sex or sick' decision. In order to create an ecologically relevant experimental situation wherein female lizards would need to invest in the healing process, wounds on their dorsal surfaces were created through a controlled surgical procedure known as cutaneous biopsy.[58] Control females went through part of the surgical process, being anesthetised without having a wound inflicted. The lizards were then placed in one of three feeding treatments: unlimited food, moderate food, or no food at all.

Reproductive investment is easily measured in these organisms, using the size and number of egg follicles in the female reproductive tract and the level of reproductive hormones in the blood. It was found that wounded females with access to unlimited levels of food resources were able to sustain both healing and reproductive efforts, but when food resources were limited, females had a lower investment in reproduction. Females in the group with no food provided had no investment at all. It seems that the tree lizard females make an appropriate tradeoff between healing and reproduction based on the resources available for healing. Because healing means survival, it supersedes reproduction when both processes are not possible.

In order to approach the same question from the angle of an increased reproductive effort, the researchers undertook a different set of experiments. Female lizards were injected with follicle stimulating hormone (FSH) in a form of lizard fertility treatments.[59] The effect of the hormone injections was an increase in the number and size of egg follicles produced by the females. Females were forced to put more resources into reproduction than they might have otherwise. Experimental females and controls that were not injected with FSH were subjected to cutaneous biopsies and placed on the moderate food diet mentioned above, which is nutritionally adequate.

Wound healing was much slower in the females injected with FSH, again highlighting the tradeoff between reproductive investment and wound healing, since both could not be done sufficiently on the moderate diet. Individuals with gaping wounds clearly are more susceptible to attacks from conspecifics and predators and to bacterial or fungal infections, so maintaining the balance between reproduction and wound healing is crucial for female tree lizards.

The natural healing process requires an increased energy investment whether you're human or any other animal, and in the cruel natural world, the required energy must come at the cost of something else. Female tree lizards cannot rely on a visit to the nearest medical clinic for a quick fix for their wounds, and an ailment or injury can result in a decrease in biological fitness if they are unable to reproduce as a result. Not only can humans easily obtain additional resources to help our bodies heal from a wide variety of ailments, but we can do so for the most part without compromising our reproductive plans. As 'unnatural' as our behaviour may seem at times, it is certainly a whole lot nicer to be a human female who can reproduce at will despite illnesses than to be a female tree lizard forced to make a compromise based on a lack of choices.

A Change of Venue

The symptoms of some human ailments seem to improve with a change of venue. While backpacking in Australia many years ago, I met a fellow who had suffered a major gash in his thigh while adventuring in Darwin, located in the far north near the equator. After a few months the wound had not healed, and the doctors concluded that he needed to relocate to a less humid environment. Once he had moved, the wound healed much more quickly. The environmental conditions in Darwin simply didn't support his body's healing of the wound.

I now live in the province of British Columbia (BC) in Canada, where it is lush and gorgeous but it rains a lot. I remember reading about a family from sunny Cape Town, South Africa, who immigrated to Prince Rupert, BC, where it rains for about 11 months of the year. Why would they do this? Several members of the family had a rare genetic disorder that resulted in a dramatic sensitivity to sunlight. Where better to live than one of the greyest cities in the world if you're allergic to the sun? Advances in human medical knowledge allow for such simple solutions to complex medical problems. Why suffer more than you have to if all it takes is a change of venue to cure what ails you?

We humans aren't the only animals to figure out that a change of habitat might benefit those suffering from certain ailments. The salmon louse (*Lepeophtheirus salmonis*) is a common parasite of many salmonid fishes. This copepod attaches itself to the external surfaces of juvenile fish and is responsible for erosion of the dermal tissue and for draining blood (energy) directly from the infected fish. A heavy load of lice (commonly called sea lice) can have dramatic negative effects on the ability of the fish to survive and reproduce. Biologists studying this relationship have noted that fish with significant infections return from the ocean to freshwater much earlier than those without.

Experimentally infected sea trout (*Salmo trutta*) smolts were found to have a high rate of return to estuarine areas from where they were released compared to control smolts that were not infected.[60] Several salmon species such as pink and coho (*Oncorhynchus gorbuscha* and *O. kisutch* respectively) with substantial infections have been observed returning to freshwater at an earlier stage in their life history (and consequently at a much smaller size) than they would otherwise. There may be specific advantages to the infected fish for doing so: sea lice don't survive much longer than one or two weeks out of the ocean. The change of salinity between salty sea and freshwater river is too much for their physiology to handle. In addition, the sea lice are unable to change hosts in freshwater,[61] making their demise in this environment more likely while at the same time increasing the survival abilities of the salmon that return there early. In laboratory experiments designed to determine the environmental salinity preferences of infected vs. uninfected pink salmon, researchers found disparate preferences between the two.[62]

When salmon could choose between a freshwater and a saltwater habitat, infected fish more often selected the freshwater habitat. The survival benefits realized by the salmon due to the downfall of the sea lice are reinforced by an indirect advantage once the infection has cleared. The osmoregulatory abilities of the fish are negatively affected by the lice. Living in the saltwater habitat is extremely costly in an energetic sense, because osmoregulating is a costly process in the more saline environment. By moving to freshwater, the salmon decrease their levels of lice infection and move to an environment where osmoregulating is much easier.

It seems simple: If you're a salmon infected with sea lice, move to freshwater. However, nothing in the world of biology is without its costs. This change of venue may benefit the salmon in terms of infection and osmoregulation, but this anadromous fish normally spends most of its life in a salt water environment where it can grow to a much larger size, since salt water contains a whole lot more food. It is possible for a small fish to survive, but with a higher risk of predation and a significantly reduced fecundity. The smaller you are, the less space there is for egg or sperm production. So while it may initially seem like an obvious choice, there are significant costs to the salmon for an early move to freshwater. As with most choices animals make, the costs and benefits must be considered in terms of the ultimate goals of survival and reproduction.

Back to our human examples—the fellow who moved across the country and the family who immigrated across the world. Both parties incurred significant costs for protecting their health, since the ability to obtain resources for oneself and one's family is initially limited upon changing one's place of residence, not to mention the cost of the move in the first place. However, like the salmon, the human travellers made the choice after weighing the costs and the benefits. You cannot hope to survive and reproduce if your health is significantly compromised in your current location.

Chapter 3

Dangers, Dramas, and Delights!

Protect the Fortress

Our homes are generally thought of as our personal space to relax and feel protected from the rest of the hostile world. *Homo sapiens* select homes based on criteria that are important to each individual, such as location and size, but certain attributes of a home are common to most of us. It is vital to have some form of security in order to ensure the safety and survival of one's most valuable assets: self, mate, and offspring. This could mean locks on the windows and doors, a guard dog, or a home security system. Although humans aren't plagued by predators in the traditional sense, there are plenty of members of our own species who prey on unlucky victims in poorly protected environments.

We humans are certainly not alone in our quest for self-protection at home; many organisms deal with the threat of intruders on a daily basis. Whether it's a mansion, burrow, cave, or nest, protecting the inhabitants is a critical function of any dwelling. Australian fiddler crabs (*Uca* species) live in burrows that they dig underground on large intertidal sand flats. The entrance to the burrow is protected by a circular chimney-like structure made of sand and mud. Constructing these chimneys represents a considerable effort, especially since it needs to be repeated each day because incoming tidal waters destroy any chimneys that are below the high tide line. In one species (*Uca formosensis*) it has been shown that the only crabs to construct chimneys are males who have recently acquired a mate. Once a female has been attracted to his burrow, he must expand it in order to accommodate the needs of his new partner. The chimneys are thought to provide visual cover to the male while he builds the expansion.[63]

In another fiddler crab species (*Uca capricornis*), both males and females build chimneys around their burrows. The chimneys are effective at deterring intruders, since the burrows are less likely to be found when the openings are concealed.[64] Although members of both sexes have been observed building chimneys, females are more likely to make the effort to construct this added precaution. Biologists hypothesize that females are less likely to be successful in fending off male intruders, so having the chimney represents an important addition to their refuge. Having lived alone as a single woman for a number of years, I can certainly understand the extra effort by the female fiddlers in the name of increased security.

Common waxbills (*Estrilda astrild*) are ground-dwelling birds that live throughout much of sub-Saharan Africa. The lengthy developmental phase of their nestlings, coupled with the fact that nests are located on the ground, makes them an obvious target for predation by rodents and snakes. However, this small bird has come up with an effective strategy to protect against security breaches: They cover their homes with carnivore poop, formally known as scat. The scat used by the waxbills comes from large African wildcats, and it contains the undigested remains of vertebrate prey. Additional cat scat is placed inside, outside, and on top of the nest every day, even though it provides no structural integrity. Instead, its function is to serve as a deterrent to small predators, who avoid the nests out of fear of larger predators, and to large predators, who often cannot detect the birds through the scent of the scat.[65] Scat loses its scent over time, so the birds must apply it repeatedly to preserve its defensive capability.

It seems as if humans have it pretty good on the home-defence front. Once that alarm system is installed or the guard dog is at his post, we generally consider ourselves to be safe from predators and intruders. We can take comfort in knowing that we don't have to rebuild our security systems each day like the fiddler crabs, and especially that we don't have to slather our homes with crap in order to keep them safe. Although animals have evolved simple and effective ways to protect their homes and their families, those developed by *Homo sapiens* require comparatively little effort. I'll vote human on this one.

Moving On Up

For members of our species, moving is exciting and stressful at the same time. Finding and being able to obtain a home that meets the needs of a growing family requires careful consideration and a lot of work. One person's discard is another person's dream: as you move out of that one-bedroom apartment into a three-bedroom condo, the owner of the condo may be moving into a five-bedroom house. Another individual, probably a first-time buyer, is happy to take over your small apartment. This kind of property exchange represents a complex process that involves several unrelated individuals or groups and can ultimately be beneficial for them all simultaneously. All the participants are happy to occupy the vacancy left by the previous tenants. Vacancy chain theory, the process of sequentially distributing resources (houses in this case) across multiple individuals,[66] occurs in our species as well as in several others. After all, there can be serious consequences for one's health and survival from living in quarters that are too cramped. Spread of disease and detection by predators are two ways survival can be jeopardized by living in a home that is too small.

Many animals build their own residences out of available materials (birds use twigs to build nests, snails secrete shells from their mantles, spiders spin webs from silk glands), making their home construction independent from conspecifics nearby. If a larger home is necessary, grab some supplies and make it happen. But other animals rely on the availability of homes in their surrounding environments the same way humans do. Clownfish that inhabit anemones, birds that nest in tree cavities, and hermit crabs that inhabit abandoned snail shells are all examples of animals that make their homes out of discrete, reusable resources that are limited in use to one individual or family at a time. Members of such species therefore stand to benefit if a new home becomes available.

Researchers examining this phenomenon in the terrestrial hermit crab (*Coenobita clypeatus*) in Belize found that there were benefits to be gained for several individuals along a vacancy chain when a new shell was introduced to a group.[67] Crabs ranging in size from large to small were placed in experimental arenas in groups of eight. A new shell that approximated the size of the largest crab was added to the arena, and the crabs were left there for twenty-four hours to proceed with shell inspections and possible shell-switching. At the end of the trials, the crabs that had changed shells were measured for differences in shell crowding between their old and new homes. Overcrowding represents a considerable survival threat for a hermit crab, because an inability to fully retract into its shell leaves it vulnerable to predators.

It was found that an average of 3.2 crabs in each group switched shells and that a whopping 89 per cent of those that switched shells gained a significant reduction in shell crowding. In other words, 3.2 individuals benefitted (increased their ability to survive) from the addition of a single resource to the arena, a phenomenon termed the chain multiplier effect,[68] where one individual stands to move on up the housing ladder by virtue of the fact that others have done so. However, the vacancy chains observed in the hermit crab population were abruptly terminated if one crab had a damaged shell. This makes sense in the human world too: damaged homes require a lot more attention and sell for a much lower price, if at all. Where we might request a home inspection before finalizing an offer to purchase, crabs also consider the qualities of a new home before moving in. The size, weight, internal volume, and amount of damage to the shell are all critical to the survival of the crabs living in them, and therefore shell selection is an important process for this and other hermit crab species.[69]

The dynamics of home selection in the terrestrial hermit crabs studied here have definite parallels to the struggles faced by *Homo sapiens* when it comes to finding and obtaining a new home. Vacancy chains can have reverberating benefits down the property ladder in both species. However, if a dwelling is substandard or damaged in some way, the benefits of a vacancy chain come to an abrupt halt. We all look for high-quality, undamaged places to live, especially when threats to our well-being exist because we've outgrown our current residence. A major difference between us and the hermit crabs as we move up the property ladder is the human concept of finance. Hermit crabs have no need for mortgages, loans, or banking, nor any need to budget the financial cost of the new home. Once an individual crab has out-competed others for a certain shell, it is that individual's home for the taking. In the human world, out-competing others for a home generally translates to making a better offer. In the hermit crab world, out-competing others means that an individual physically fought for it and won. Perhaps it's a little on the violent side, but at least it's simple—and mortgage-free!

Trick or Treat?

Obtaining the resources that one requires in order to successfully survive and reproduce is a goal that some members of our species achieve dishonestly, despite the fact that others may be harmed in the process. How do they get away with it? Criminals utilize the powers of trickery and disguise in a variety of ways, from making themselves difficult to detect (dark clothes, face masks) to practicing disguise and deception to steal your hard-earned savings from right under your nose. Although at times I feel this is an unfortunate aspect of human nature, it turns out that such trickery is not limited to the calculating ways of the *Homo sapiens*. Scams and deception are commonplace in the animal kingdom, and there are serious consequences to biological fitness for individuals who fall victim to such schemes. Outwitting your opponent can be an extremely important aspect of survival, for both the tricksters and their victims.

Cleaning stations are areas on coral reefs where specific types of fish will perform the task of removing parasites or other detritus from client fish (see Keep it Clean!). The system works quite well and is mutually beneficial: the cleaners (such as the bluestreak cleaner fish *Labroides dimidiatus*) get easy meals that come right to their doorstep, and the clients (a wide variety of reef fish) are able to rid themselves of troublesome parasites and have their abrasions attended to.

Such a happy arrangement does not come without dangers in the form of trickery. *Aggressive mimics* are organisms that resemble a nonthreatening or attractive organism for the purpose of gaining benefits.[70] Bluestriped fang blennies (*Plagiotremus rhinorhynchus*) are aggressive mimics of juvenile bluestreak cleaner fish. Instead of removing parasites from client fish as a normal cleaner would do, the fang blennies remove a chunk of flesh. They can trick their prey into coming right over to them before launching their attack. All is not lost for the client fish, however. Researchers studying this phenomenon have noted that the mimic attacks become less successful over time, as the client fish learn to avoid both the mimics and the areas in which they work.[71] Cleaning stations are plentiful in a coral reef, and stations that have aggressive mimics present have been shown to receive fewer visitors than those that do not.[72] The trick, as with any kind of behaviour, is moderation. Once there are too many mimics out there, victims become wary and suspicious.

Predators aren't the only ones that resort to trickery to gain an advantage. Biologists assume that selection pressure is actually greater on prey species than it is on predators, because the prey has much more to lose. The life-dinner principle[73] describes how the situation differs between predators and prey. Predators may have to skip a meal if their disguise doesn't work, but prey may lose their lives, which provides a strong incentive to make sure they can outwit their opponents.

California ground squirrels (*Spermophilus beecheyi*) have been preyed upon by Northern Pacific rattlesnakes (*Crotalus oreganus*) for millions of years. The squirrels have evolved a unique method of disguising themselves from their main predators: they dress up like them. The squirrels masquerade as rattlesnakes by chewing on shed snakeskins and vigorously licking the scent onto their own bodies.[74] By covering themselves with eau de rattlesnake, the squirrels mask their own natural scent and protect themselves from their foremost predator. Rattlesnakes are more attracted to the pure scent of the ground squirrel than they are to the scent of ground squirrel mixed with rattlesnake.

Tricksters are abundant in all societies, and our species is no exception. Tricks can be designed to gain resources in a dishonest way, like the disguised predator, or they can be used to protect resources by fooling those who want to take them, like the disguised prey. Luckily for humans, those who are trying to trick us are usually doing so to secure resources rather than to eat us for dinner. For most other animals, the stakes of such trickery are much higher. Selection pressure on prey organisms whose lives are on the line has resulted in a myriad of strategies that serve the sole purpose of ensuring their survival. On the other hand, to keep up with the tricks of their prey, predators have developed ways to detect disguises or wear disguises of their own. This illustrates wonderfully the beauty of evolution, although it strikes me as disconcerting that so much energy is spent on the evolution of the perfect trickster.

Dress Up or Dress Down?

Sometimes surviving is all about blending in: being a faceless member of a crowd or going about your business without anyone noticing. Roaming around a dangerous neighbourhood at night isn't a great idea to begin with, but it can be made much worse if you wear a flamboyant outfit and dramatically announce your presence. In such a situation, maintaining a low profile is a much smarter strategy for avoiding danger. For most members of the animal kingdom, blending in is a strategy employed to avoid being detected by predators. Although humans are not likely to be targeted by a predator in the same way as other animals, dangers exist from within our species that justify the blending-in strategy as well. We have access to appropriate decorations such as dark clothing or eyeglasses, to make this strategy effective.

While camouflaging oneself makes the most sense in some social situations, others require a different approach. There are times when we'd like to be noticed, especially by the opposite sex. If we've got mating in mind, the decorations we select should reflect the need to make potential partners aware of our existence. After all, Mr. or Ms. Right is more likely to take notice if we stand out from the others around us. Many organisms in the animal kingdom are physically camouflaged and cannot change their appearance in alternate social situations, but others, like humans, are able to change their external decorations depending on the costs and benefits of either standing out or fitting in.

Consider the situation of the rock ptarmigan (*Lagopus mutus*) living on the plains of Arctic North America. The white plumage of both males and females disguises them well through the winter months when snow covers the ground, but once the snow melts, these bright white birds are clearly visible to predators against their new brown background. Consequently, females immediately moult their white feathers and take up a new summer coat of brown feathers that allows them to blend into their surroundings. Males, however, remain white for a number of weeks, rendering them susceptible to predation. Dressing in white represents a major tradeoff for the males, since their white plumage is a sexually selected trait designed to win them an appropriate mate. They don't want to moult early and risk not finding that special someone, but they also don't want to fall into the clutches of a predator before being reproductively successful.

A 17-year study of these birds in the Arctic confirmed that males get right to work creating their disguise until they are able to moult into their summer coats.[75] As soon as mating is complete, they begin actively soiling their bright white plumage with copious amounts of dirt and mud, reducing their conspicuousness by as much as sixfold against the drab background. They become extremely dirty in a short time, quickly creating an effective deterrent against visual predators. During the course of the mating period, male ptarmigans retain the ability to clean themselves back to their showy white selves if a new mating opportunity presents itself, but until an eligible bachelorette comes on the scene, they remain incognito.

Spider crabs also maximize the power of external decoration. Members of the family Majidae are known as the true decorator crabs. These crustaceans cover their exoskeletons with various species of algae, hydroids, sponges, and a number of other creatures. Using specialized Velcro-like bristles on their external surfaces they can apply adornments wherever they like. The function of such elaborate decoration is to avoid predation, for it provides an effective camouflage against the colourful background of the substrate that the crabs inhabit in the same way as the dirt and mud decorations of the male ptarmigans.

There are physiological costs associated with carrying around an abundance of other organisms, however. Wearing such a disguise can mean a tradeoff between predation risk and energy loss. Generally, spider crab species that remain small as adults (and therefore have a lifelong risk of predation) remain decorated throughout their lives, but species that grow to a larger size as adults tend to decorate only as juveniles.[76] Once the larger species are large enough to be less vulnerable to predation, disguises are shed and energy is made available for other purposes, such as finding a mate.

These two creatures don their disguises in opposite ways. Male ptarmigans are at first showy in order to find a mate, and then they camouflage in order to avoid being detected by predators. Decorator crabs are showy as juveniles or small adults in order to be camouflaged, but they become more visible when predation risk is lower and energy can be turned towards reproduction.

Dressing up or dressing down is a complex business in the animal kingdom, but how about *Homo sapiens*? Our species is lucky in that changing our decorations is a simple process. Unlike the painstaking efforts required by the ptarmigans and the crabs, most of us need only look in our closets for adornments appropriate to the occasion. In addition, while fitting into certain social situations can be critically important, most of the changes of decoration made by members of our species have little to do with immediate survival or reproduction. How easy it is for us to forget that the rest of the animal kingdom doesn't have such freedom!

Pay Attention or Pay the Price!

Vigilance is an important attribute in today's society. Tragedies, both accidental and deliberately imposed, are most likely to occur when we aren't paying enough attention. Unfortunate as the consequences for survival may be, people seem to make a point of not being attentive. We talk on cell phones while driving; we don't check the locks on our doors at night. *Homo sapiens* in the western world are generally so overburdened with daily tasks (many of which have nothing to do with biological fitness) that the basic rules of vigilance sometimes escape us. It's when we're occupied with other tasks and not paying enough attention to the actions of others that 'accidents' can happen. Predators are often on the prowl for those of us who are inattentive. Career criminals study the habits of their victims, understanding when they come and go and when they are more vulnerable. They capitalize on our mistakes, such as leaving the house without locking a window or working late and walking through a dark parking lot alone. It is at these moments that they seize their targeted prey. A little extra caution goes a long way in such situations, but as I mentioned before, we often multitask to such an extent that paying attention isn't our top priority.

We are certainly not the only species faced with such a conundrum. Many organisms balance a multitude of tasks in their day-to-day existence just as *Homo sapiens* does. Looking after offspring, maintaining the home, and finding and preparing food are just a few of the responsibilities facing most members of the animal kingdom. Unfortunately, just like the fallible human, other animals are unable to invest sufficient effort in avoiding dangers when their brains are occupied with difficult tasks.

Blue jays (*Cyanocitta cristata*), for example, have a limited attentiveness to predation risk when faced with a difficult foraging situation.[77] Biologists altered foraging arenas to reflect easy and difficult foraging circumstances by providing either a small or a large number of nontarget (background) items in the arenas along with the prey. When more effort was required to correctly identify a prey item, the birds were less able to respond to the peripheral disturbances (simulated predation) induced by the researchers in the arenas. In other words, when it took more brainpower for the blue jays to detect their food, they were unable to put in as much effort on other important tasks such as being vigilant against predators.

Predators are influenced by this circumstantial lack of cognitive ability in their prey. Jumping spiders of the genus Portia have been shown to exploit the limited vigilance of their victims by exhibiting opportunistic smokescreen behavior.[78] They take advantage of times when their prey is otherwise distracted (the smokescreen) to apprehend them easily. The Australian jumping spider (*Portia fimbriata*) preys on a web-building spider (*Zosis genicularis*). The jumping spider has well-developed visual capabilities and can utilize optical cues to determine the status of a potential target. Unlike the complex detection system of the predator, the web-building prey spider has poorly developed visual capabilities and relies instead on web vibrations to detect intruders.

Jumping spiders increase their chances of a successful apprehension by attacking the web-building spiders when the prey is consumed with wrapping up an insect that has been caught in its web. Not only is the attention of the web spider diverted from vigilance by the task of wrapping its own prey, the wrapping behaviour creates movement and vibration in the web that masks the ability of the web spider to detect the approaching predator. By waiting for the attention of the web spider to be directed elsewhere, the jumping spider acquires its dinner easily.

Although biologists often assume that animals make an optimal tradeoff between meeting their basic needs, such as obtaining food and remaining on guard for predators, there is no avoiding the fact that some tasks are difficult to do. It isn't easy to detect food items in a complex environment or wrap newly captured prey. These actions are necessary to survive, and they require a good deal of attention. In this way *Homo sapiens* is much the same as any other creature that comes to harm through a lack of attentiveness, but the underlying reasons for the inattention are altogether different.

The blue jays and the web-building spiders are busy capturing their meals and preparing them for consumption, both necessary components of survival. We humans, on the other hand, often don't pay attention to dangerous situations because we're consumed with aspects of our lives that have little to do with our immediate survival. Predators are watching whether you're a human or any other animal, so wake up and pay attention!

Slackers among Us!

Having run a small company for several years of my life, I am well aware that the sum is only as good as the parts. The importance of having efficient employees cannot be overemphasized, as things can get quite unproductive if people aren't pulling their weight. Slackers, individuals who don't work to their full potential, create a greater workload for others and decrease efficiency. Slackers are an unfortunate reality of many workplaces. From a selfish standpoint, it makes sense to be a slacker: individuals who can successfully mask their laziness are able to save their energy for other, presumably more selfish, purposes. Slackers fare particularly well if they can hide behind the efforts of workers who are efficient and successful at performing more than their fair share of work. The overachiever is another common type of employee, the person who prefers to take on a greater-than-average workload. For the boss and for the company, this is a very good thing, but the presence of overachievers can mask the identity of the slackers by compensating for their laziness. What's a boss to do? One can look to the animal kingdom for advice from the most efficient companies in the world.

Colonial insects operate much like large companies. Ants, bees, wasps, and other colonial insects are enormously successful in the animal kingdom, totalling up to 75 per cent of the world's insect biomass.[79] One of the reasons often cited for their success is the fact that individuals in the colony specialize in specific tasks, such as structural maintenance, care of young, food gathering, and defence. This increases the level of efficiency at which all tasks are performed, like having specialized employees in the accounting department and others in public relations.

The insect world is not without its slackers. In colonies of the ant *Temnothorax albipennis*, a high proportion of individuals actually perform little or no work.[80] When it comes to certain tasks, elites or key individuals are relied upon to do a great deal of work, whereas others perform relatively little.[81] This imbalance in workload is particularly evident during collective emigrations, where the entire ant colony vacates one site for another more desirable location. A great deal of work is required in order to move the entire colony. One would assume that the most efficient method would be to distribute the work among all capable individuals, but the opposite is the case. There is a major skew in the amount of work undertaken by individuals during the emigrations, with some overachievers doing a disproportionately large amount of work and slackers doing none at all. In one observed case, a single individual was responsible for transporting 57 per cent of the nest items to the new site.

Both ants and humans who get a free ride are compromising the fitness of those around them. However, for the humans at least, the overachievers generally have the last laugh. By working hard and moving up the ranks, the overachieving *Homo sapiens* will likely become a manager in the future and can fire the slackers who compromise the efficiency of the organization.

For the ants, the story is different. Biologists argue that the experience of the emigration overachievers allows them to transport materials more efficiently than most other individuals. Their experience in locating materials and optimizing a path to the new nesting site means that resources can be moved more quickly if inexperienced individuals do not help. In other words, the slackers are slacking because overall it's better for the colony to make its move quickly and efficiently, and these individuals would slow the process if they tried to help. The overachievers are doing the extra work for the good of the colony, without the guarantee of a promotion or a corner office. Not exactly the most sound logic, if you ask me.

This conundrum of the ant-slackers slacking for the good of the colony may be unique to the task of emigration. Other daily tasks such as food gathering, brood rearing, and nest maintenance should not be prone to the same rationale for not helping out. Let's hope not. The ant colonies, just like the major companies and corporations of *Homo sapiens*, would be in big trouble if a misguided policy of tolerating slackers for the greater good was applied to the work required for survival.

Cat Fight!

It is rare to witness a full-on female-female physical altercation in the human species. It does happen, but not nearly to the extent to which it occurs between males. Female-female fights are not pretty. They involve hair pulling, scratching, pinching and other tactics that generally reiterate their status as the sex that isn't supposed to be so brutal. Males, on the other hand, undertake physical battles on a regular basis and use the traditional tactics of punching and kicking. Not only are there physical differences between males and females with respect to confrontations, there are often emotional differences with respect to how fights begin in the first place. Females are more likely to fight with a high degree of emotional involvement, whereas males are more likely to fight with just the physical contest in mind. So why do such fights within species occur at all?

When it comes right down to it, physical confrontation between members of the same species, human and otherwise, is most likely to occur due to a resource limitation. Whether the coveted resources are critical to survival, like food and shelter, or to reproduction, like an appropriate mate, conflicts occur within species all the time. *Homo sapiens* does not appear to be the only animal that exhibits a difference in fighting tactics between the sexes.

The Texas cichlid (*Herichthys cyanoguttatum*) is a species of freshwater fish that has been studied extensively with respect to its aggressive behaviour. Fighting between members of the same species (intraspecific confrontation) occurs both between males and between females. However, there are striking differences with respect to the manner in which the fighting is initiated and carried out, depending on the sex of the combatants.[82] The outcomes of male-male fights are almost always determined by the resource holding potential (RHP) of the specific males involved.

Essentially, the capability of a certain male to obtain and maintain his hold on a resource determines his potential for victory. The most obvious component of a male's RHP is his size. A large male has a natural advantage during physical conflicts because he can intimidate and overpower smaller competitors due to his greater RHP. The outcomes of female-female fights, however, have nothing to do with size. The level of aggression exhibited during what we call catfights is related to the resource payoff value (RPV) of the item involved. Unlike the RHP, the RPV has to do with how much an individual will benefit from winning the resource. A female who is close to spawning will fight harder for a spawning shelter regardless of her size because she needs to rear her young in a safe place. If a female is not close to spawning, she will not aggressively fight for a shelter. Overall, the fighting behaviour between the cichlid sexes suggests that males follow a *Can you win it?* strategy, whereas females follow a *Do you want to win it?* strategy. Sound familiar?

The yellow-rumped cacique (*Cacicus cela*), a nest-inhabiting bird studied in the forests of Peru, exhibits a similar disparate pattern between the sexes. Fights between females are usually for nesting sites and/or nesting materials, and behavioural observations have revealed that females fight more intensely for partially constructed nests, which have a higher RPV value, than for nesting sites with no previous construction.[83] Such observations confirm that the value of the resource to the individual plays a significant role in the intensity of female aggression. On the other hand, escalation of fights between male yellow-rumped caciques was found (as with the cichlids) to be more related to the relative size and RHP of the contenders. Fights between similar-sized birds tended to last longer and to be more aggressive than encounters between birds of significantly different sizes.

So when it comes to catfights in the animal kingdom, common themes emerge in the behaviours of females, from fish and birds to humans. Females will fight more intensely for resources that are important to them, such as an appropriate mate or a safe place to live. Such resources directly contribute to their biological fitness and are therefore worth fighting for. However, if a resource is not required at a particular time, females will not waste their energy in a battle to obtain it. How very sensible! Males, on the other hand, humans included, will fight based on their ability to win, regardless of the importance of the resource involved. Ahhh, the universal power of testosterone.

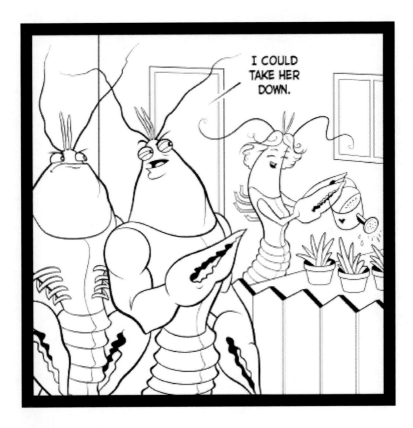

Never Hit a Girl!

Although chivalry may be dying out in many human societies, there are certain things that most of us *Homo sapiens* respect as social no-nos. One of them is a male using physical aggression against a female, which is frowned upon to such an extent that it rarely occurs. No matter how much a man might want the resources she holds (even if it's the last slice of pizza in the box) there is never a good excuse to use physical aggression in order to get it. Based on the physical differences between males and females of our species, the former have a clear advantage over the latter and therefore such conflict is usually both unfair and unnecessary. However, for most organisms in the animal kingdom, obtaining adequate food or shelter is not simply a matter of luxury as it often is with the *Homo sapiens*—it's a matter of life and death. It is therefore conceivable that when certain resources are critical to survival, all rules of politically correct engagement are off. Most animals duke it out in order to obtain required resources with little regard to the sex of their competitor.

One example is the hermit crab (*Pagurus* species), an organism in which members of both sexes have a vital requirement for shelter. Since these crustaceans do not have a hardened abdomen like most other crabs, they require the protection of a gastropod (snail) shell in which to live. These homes are a hot commodity in the hermit-crab market, and having one large enough for an individual to completely retract into is imperative in order to avoid predation (see Moving On Up). Conflict over shell ownership is common in several hermit crab species, and such conflict sees no distinction between the sexes. Individuals that do not currently inhabit a suitable shell will attempt to steal a shell from another, leading to a high potential for aggressive encounters, regardless of whether one is male or female. There are two players in any shell-heist scenario: the defender is the crab who currently inhabits a given shell, and the attacker is the crab who is attempting to evict the defender. An attacker will approach a defender and attempt to take over the shell by manipulating it with the abdominal muscles and walking legs and swinging it up against its own unsuitable shell in a series of raps.[84] There may be several bouts of rapping, after which the defender is evicted from its shell or the attacker gives up. Attackers who are more aggressive in their rapping behaviour and who take only short pauses between raps are more likely to win encounters than those who are less aggressive.

Investigations of similar-sized male and female hermit crabs (*Pagurus bernhardus*) in defending and attacking roles show distinct differences between their abilities.[85] Males are most likely to initiate an attack, whether against males or females. Males also appear to have an overall advantage in both attacker and the defender roles, more often successfully obtaining the coveted shell or defending their own shell. Although females appear to fight harder than males for access to new shells by leaving shorter intervals between rapping bouts, males are nevertheless more likely to be successful on either side of a fight.

Such a male-dominated success rate is not exhibited in shelter competitions of the red swamp crayfish (*Procambarus clarkii*). Securing and maintaining a shelter is extremely important for survival, most specifically for maternal females (who need a place to raise their young) and adult males (who are particularly susceptible to predation by fish or birds). Although adult male crayfish are able to defend their shelters against *nonmaternal* females, experimental evidence shows they are not successful at defending them against *maternal* females.[86] (If mama's got babies to look after, watch out!) When females are defending their shelters, the story is similar: males can always evict a nonmaternal female, but they cannot evict a maternal female, despite repeated attempts to do so. When it comes to intersexual conflicts in this species, maternal females are tough to beat. Biologically speaking, this makes a lot of sense. A maternal female with a clutch of offspring has much more to lose in terms of biological fitness if she loses her home.

In the animal kingdom, chivalry is not only dead, it was likely never alive in the first place. When males and females have overlapping requirements for resources that will impact their ability to survive, both will fight aggressively in order to obtain them. If this means physically overpowering a female and turfing her from her home, then so be it. Although in this day and age a human male may be less likely to hold a door or give up his place in line for an unfamiliar female, at least he is a little more civilized than the hermit crab or the crayfish when it comes to his treatment of females. We female humans are unlikely to be physically overpowered by a male competing for a similar resource. In the arthropods' defence, it can be noted that the resources for which *we* compete have little to do with our immediate survival; perhaps if they did, *Homo sapiens* ladies would not have it so easy.

The Power of Peers

Members of our species rely heavily on others to learn about resource value in matters such as choosing food, attracting a mate, or finding a place to live. When we are young juveniles, parents and close kin play a significant role in our learning processes, but once children fledge the nest and are on their own in the real world, they must be able to make such choices without the help of Mom and Dad. So what can we do to increase our level of success when it comes to selecting appropriate resources? We can observe other humans, of course.

When selecting an appropriate habitat to settle and reproduce, one might investigate different districts, observing others who have chosen to live there and assessing the amenities in the area. Chances are that if there are a lot of young families in a neighbourhood, it will offer resources favourable to folks in that demographic. Inadvertent social information can be passed between individuals in various ways. The mere presence of individuals in an area can be a signal that there is something there worth getting, but many organisms take public information use to the next level.

Noting the behavioural decisions of others with respect to habitat selection can go a long way towards improving the fitness of the observer with very little work required. If we see there are two ways to gain information about a resource—by sampling it yourself or by obtaining it from others—we should assume that an individual's biological fitness is maximized by the action that has the lowest energetic cost.[87] When collecting public information is easy, such as by observing the actions of others, such a method should be preferred, for it is generally a more reliable indicator of the potential fitness benefits to be gained. However, if collecting public information is difficult, perhaps because others are not easily observed, one would expect that decisions will be made based on self-sampling of potential environments.

We humans may assume that we are the only species making complex decisions based on material as well as social parameters, but this is far from the case. In the animal kingdom, many species use publicly shared information to make appropriate decisions with respect to resource use, and this can have a dramatic impact on their ability to survive. Black throated blue-warblers (*Dendroica caerulescens*) are North American songbirds that are specialized with respect to their nest-site selection. Areas with dense shrubs and mature trees are generally preferred over other habitats. The structure of the surrounding vegetation has traditionally been assumed to be the most important factor in the habitat choice of this and other bird species.[88] However, the importance of social cues (as opposed to vegetation structure) in habitat selection by these birds is not to be underestimated. During the breeding season, social signals—sounds of successful reproduction, such as postmating songs and sounds of chicks begging for food—are abundant in certain habitats. These social signals convey information to a potential 'home buyer' about the survival and reproductive benefits to be gained from living there.

Field experimentation on habitat selection in prospecting warblers has demonstrated that the social signals are, in fact, the *strongest* factor influencing habitat selection.[89] Researchers played audio signals of postmating songs and chicks begging for food in areas that had little shrub cover or mature trees (resource poor environments) in order to test the influence of social cues. They found that individual warblers overwhelmingly selected to live in the 'inappropriate' habitats over forested areas nearby. In other words, the effect of the social cues on habitat selection is strong enough to result in recruitment to areas that are completely unsuitable!

Homo sapiens is also capable of being influenced by social information to an extent that might overshadow the appropriateness of a decision based on food or habitat. Eating at a flashy restaurant, regardless of the kind of food available there, or choosing to share a tiny apartment in a hip area of town are decisions that can be heavily influenced by social learning. They don't necessarily translate to a higher level of biological fitness. It seems that our common sense can be momentarily altered so that social information takes precedence over our basic needs. Although not surprising for the biologically confused *Homo sapiens*, it is interesting that the decisions of the warblers are also influenced that way. The individual birds that selected the resource-poor environment gave up their personal point of reference, even to the point where there was no food in sight, in favour of the fact that others seemed to have done well there. So humans aren't the only animals who are subject to the power of social influence, which makes me feel a little more at home in the animal kingdom.

The major difference is this: in the 'real' world, reproductively successful warblers would not be singing reproductively successful songs in environments where they had not been reproductively successful. The researchers created this unnatural setting in order to determine the importance of social cues to the birds; such a situation would not occur naturally. Warblers, unlike humans, do not hang out in unhealthy places just because it's the cool thing to do.

Take Out the Trash!

All species produce waste. Human societies worldwide create an immense amount of garbage, both biodegradable and not. The amount created by a particular society is generally proportional to its size, although in the western world we are guilty of producing much more than necessary with our mass consumption of overpackaged products. It is therefore imperative to develop strategies to handle the vast quantities of refuse, as unchecked buildup can result in an abundance of parasites and disease. In my neighbourhood, as in many others across the world, we have a weekly garbage pickup. Waste products are placed in bins at the foot of our driveways, reliably picked up by waste management workers, and taken to a large heap somewhere out of town, far away from where we have to look at it, smell it, or associate with it in any way. This system works quite well, as it removes both the waste and the potential for harm to our health and survival that comes along with it. Out of sight, out of mind, and we all live happily ever after. As for the animal kingdom, waste accumulation is not a problem for nomadic organisms that do not occupy one area for too long, or for water-dwelling organisms that capitalize on the power of the currents and the depths to carry their wastes away. However, for animal societies with consistent nests or colony locations, an effective strategy is imperative in order to avoid the problems associated with buildup.

Disposal of sewage, for example, represents a major waste-management issue for group-living organisms. Lacking the convenient plumbing and flushing methods developed by *Homo sapiens*, animals have come up with other effective strategies for managing their bodily wastes. Spider mites (genus *Stigmaeopsis*) are tiny arachnids (adults are between .3 and .5 mm in length) that build nests on the undersides of the leaves they eat. Most species of spider mites practise some form of waste management with respect to where they defecate. Some species build simple nests with two entrances and have a defecation site near each entrance outside the nest. The mites are able to find the locations of the defecation sites using tactile cues.[90] Some species of mites construct complex nests that are continuously being enlarged in order to house additional conspecifics. Tactile cues at nest entrances don't work as well for these mites, since new entrances are continuously being added, so chemical cues in the feces are utilized in order to find the appropriate place to defecate. Whether through tactile or chemical pathways, spider mites keep their nests clean by making sure they do their business elsewhere.

Taking it a few steps beyond a simple defecation area, leaf-cutting ant societies have elaborate and organized protocols for dealing with buildup of sewage, deceased workers, and other organics. Some species utilize underground dumps for their trash, while others have designated heaps outside their nests. Waste-management strategies documented from more than 50 colonies of leaf-cutting ants (*Atta colombica*) in Panama show that they employ a detailed set of associated guidelines with respect to garbage-handling.[91] The refuse produced by these organisms contains fungus and the bodies of deceased workers, and it is likely to become toxic to colony members if it isn't promptly removed from the immediate vicinity of the nest. Approximately ten per cent of the worker ants in these colonies are devoted to waste-management tasks, and there is a strict division of labour between foraging workers and trash workers. No task switching occurs between members of these two groups, reiterating that the potential contamination costs are high. Would you want your garbage pickup person to take a break in order to fix your lunch?

Among the waste workers are transporters, who carry trash to the heap, and heap workers, who rearrange and organize the pile for speedier decomposition. In addition to the worker detail surrounding the garbage collection, there are some aspects of the heap itself that appear to have been carefully considered. It is always located downhill from a nest entrance, which minimizes wastes accidentally entering the nests during a rainfall. Foraging trails used by the colony workers to collect foodstuffs are constructed away from the trash pile, and a whopping 92 per cent of trails used by one large colony led away from it. Although the overall trail lengths were increased by 6 per cent to compensate for avoiding the garbage dump, the extra work required to forage appears to be overruled by the need to avoid exposure to the toxic buildup.[92]

In much the same way that garbage is taken from my house to some faraway place that I have yet to visit (and probably won't), the waste-working ants transport garbage to dumps that aren't frequented by other members of the colony. The workers who transport my garbage will not be switching tasks at midday to provide me with food, and those in the ant colony won't be feeding their conspecifics either. Basic hygiene is definitely recognized in the animal kingdom—for those living in close quarters with a few thousand of their closest friends and relatives, survival depends on it.

PART II: REPRODUCE

Chapter 4

The Dance Begins

Awkwardness Across the Animal Kingdom

Puberty represents the first ticking of our biological clock, the onset of the countdown to the time when we will no longer be able to fulfill our biological destiny. While most of us are not interested in spreading our genetic material across the *Homo sapiens* landscape immediately at the onset of sexual maturation, such an important event is not to be taken lightly, lest we become unable to reproduce at some future time. For the most part, we humans don't have a way to control or predict the onset of puberty: the signs of growing up (and growing awkward) present themselves when they're good and ready, often at the most inopportune moments.

Unlike *Homo sapiens*, some animals possess the ability to control the onset of sexual maturation, and the reasons for doing so are interesting and complex. Mature female wild house mice (*Mus musculus*) have been shown to suppress the onset of puberty in juvenile females through a chemical cue in their urine.[93] Urine collected from adult females in high-density populations was found to delay puberty in other females, but urine collected from adult females in low-density populations did not. This suggests that the decreased population growth rate caused by the delayed onset of sexual maturation in juvenile females serves to regulate the size of the population when things are getting a little too crowded. Not to have their reproductive destiny truncated by a puberty-crushing diva, a chemical cue from the urine of adult males has been found to have the opposite effect. Socially dominant male mice have an increased level of pheromones in their urine that acts to stimulate puberty in juvenile females.[94] These high pheromone levels are suppressed in subordinate males and in high-density populations with multiple males, again suggesting a kind of self-regulation of the mouse population through physiological constraint. Pretty darn sophisticated, if you ask me.

The onset of sexual maturation is also controlled in other mammals. Consider the commonly observed social strategy of one alpha male living in a group with a number of females and offspring. A young male (not the alpha) faces a conundrum when it comes to sexual maturation: if he matures early, he will be subject to aggression from the dominant male and will likely be forced to forge a new group on his own. However, if he matures late, he may be limiting his spectrum of reproductive potential. The effect of the social environment on the sexual maturation of male orangutans (*Pongo pygmaeus*) is more dramatic than in any other primate. In these organisms, one dominant male in a group develops pronounced secondary sexual characteristics such as massive body size, large cheek phalanges, and a muscular throat pouch.[95] Young males fail to develop these characteristics when in close proximity to a dominant male, in a form of arrested sexual development. Despite not having the full suite of sexual characteristics, arrested males are still sexually mature and capable of siring offspring. The females however, are much more likely to be interested in copulating with the dominant male (see Sneak It In When You Can), leaving the sexually arrested male with a decreased overall reproductive success and a need to force copulations on unfortunate partners.[96] Arrested males retain the ability to ascend the puberty ladder should the opportunity arise: if they are removed from the group or if the alpha male dies, another will step up to fill in the position by developing the secondary sexual characteristics.

It seems then, that our species is challenged when it comes to altering the timing of our sexual maturation. Are we flushing all of those pheromonal cues down the toilet or is there a physiological limitation on our level of control over our reproductive status? One thing is certain: our careless use of unnatural hormone-laden chemicals and other products has resulted in some alarming changes to the timing of reproductive maturation in many human populations. Such changes, usually an earlier onset, can result in females becoming reproductively mature at an age when their bodies cannot support a pregnancy or a natural childbirth. This is the opposite of an evolutionarily sound reproductive strategy. Such changes result from the unnatural actions of our species alone, but our careless—and shameful—actions have effects that reverberate through the ecosystem to other unsuspecting organisms. The shameful wrath of the *Homo sapiens*!

Are Those Real?

Human females in the western world and elsewhere tend to be picky about their physical appearance. A visit to the beauty salon or plastic surgeon can work wonders for the appearance as well as the ego. After all, such beauty interventions are meant to function for the greater good of the species, are they not? Cosmetic alterations are thought to attract the interest of the opposite sex so that we may successfully propagate our genetic lineage. To any visually oriented organism, the appearance of a potential mate is paramount. Some members of our species will argue the importance of personality for a long-lasting union, but let's face it—what initially draws us to potential mates is how they look.

Homo sapiens is one of the many species in the animal kingdom where members of the opposite sex are assessed for appropriate physical characteristics, and partnerships are formed based on the outcome of such assessments. Physical differences among potential partners in the animal kingdom can be subtle, reflecting differences in body condition or musculature, or they can be flamboyantly obvious.

The peacock is frequently cited as an example of sexual selection at work,[97] and for good reason. Males are adorned with multiple long feathers that provide females with a breathtaking display of prowess and beauty when displayed. The number and condition of feathers during a full display are indicators of male beauty in these creatures. Females choose males that show the biggest and brightest presentation.

Feather coloration is an important aspect of male beauty in many other birds as well, but beauty is most definitely in the eye of the beholder. The eyes of many bird species (unlike those of *Homo sapiens*) can discriminate wavelengths within the ultraviolet (UV) spectrum, and it has been found that the UV nanostructure of feathers is important in female mate choice in such species as western bluebirds (*Sialia mexicana*),[98] blue tits (*Cyanistes caeruleus*),[99] and ostriches (*Struthio camelus*).[100] The higher the UV content, the more beautiful the males are considered to be.

Organisms that don't have feathers to impress potential mates possess a multitude of other physical characteristics that serve the same purpose. Male tree lizards (*Urosaurus ornatus*) can have one of several throat color morphs (blue, orange-blue, or yellow-blue, in differing proportions) that have been linked to female mate-choice. Male guppies (*Poecilia reticulata*) possess elaborate colour patterns involving blue, green, violet, yellow, orange, and black. Female guppies have been shown to prefer males that stand out from other conspecific males by showing a unique (as opposed to a redundant) color pattern.[101] Among these guppies, being beautiful means standing out from the crowd. In other organisms, the size of sexual ornaments can influence a female's decision, as evidenced by studying antlers, horns, and mandibles on organisms from dung beetles to caribou.[102] The sheer diversity of the ways beauty is manifested in the animal kingdom is mind-boggling.

The major difference between human species and most others is this: in the natural world, ornamentation and behavioural displays are accurate, honest indicators of health and fitness. It may seem that beauty is only skin deep when it comes to mate selection in sexually adorned creatures, but what is really happening is the selection of good genes. Conversely, the beauty interventions utilized by humans aren't natural and don't indicate good health or good genes. Do beauty interventions make us dishonest mates? I believe they do.

Sometimes the least desirable genes are done up in a good-looking package, leaving us in a predicament when it comes to mate selection. Beauty in the natural world is pretty straightforward: it goes hand in hand with health and fitness. Beauty in the *Homo sapiens* world is not necessarily an indicator of either one. I guess there might be something to be said for choosing a mate based on their personality after all.

Virgin Territory

For humans, losing one's virginity is a monumental event. For both males and females it is a pivotal time that signals the beginning of our sexually active, and hopefully child-rearing, lives. The sexual status of a potential partner is vastly important to some members of our species. Among males, there is generally a great deal of enticement in deflowering a virginal female. Does this have any biological meaning?

For many animal species it is important to understand the sexual status of a partner, but the reasons are different for males and females. Biologists generally assume that males want to maximize their reproductive output, making the virginity of a female quite desirable—if she hasn't mated with any others, his sperm will likely achieve paternity of offspring. Males use a variety of signals to distinguish the sexual status of a potential mate, including body coloration, auditory or olfactory cues, and behavioral signals. They then use this information to spend their reproductive energy on the most rewarding partners, namely virgins. Olfactory (chemical) cues are commonly utilized by organisms that live in water, where signals can be effectively transferred over long distances.

Male guppies (*Poecilia reticulata*) have been shown to rely on such olfactory signals to tell virgin females from mated ones. In laboratory experiments, male guppies were shown to follow virgins, nip at them, and copulate with them much more frequently than with previously mated females.[103] When all chemical cues were removed from the experimental arenas, the males no longer expressed a preference for virgin vs. mated females, indicating the importance of the chemical signal for mate detection in this species. The use of chemical signals to find a virgin female has been demonstrated for the freshwater crayfish *Orconectes quinebaugensis* as well.[104] Males were shown to be sexually attracted to virgins but unresponsive to previously mated females.

Despite the fact that virginal females may seem to be the most enticing to hopeful males from humans to crustaceans, the same cannot be said for female choice. To put it bluntly, what kind of female wants a man who doesn't know what he's doing? If he hasn't had sex yet, there may be a good reason for it. Maybe there's something wrong with his ability to procreate. Unlike males, females have limited abilities to reproduce, so they have to be choosy.

A human male could conceivably (pun intended) reproduce several times a day, but a human female requires nine months or more between successful reproductive events. A female should therefore prefer a male who can has the highest level of biological fitness. Males with a great deal of sexual experience are more likely to have a higher social status and genetic quality than males who haven't yet had the chance to copulate. It has been shown for the moth *Utetheisa ornatrix* that improving one's mating techniques in order to stimulate females more effectively can be rewarding.[105] Males with more mating experience are preferred by females, which biologists attribute to the indirect benefits females receive from mating with such males (i.e. good genes for their offspring). In addition, the courtship ritual in this moth involves a complicated set of maneuvers, so the more experience a male has in performing them the better.

In my opinion, the human species fits quite nicely into this context. Although a human male may not have paternity in mind when he attempts to copulate with a virgin female, one can probably generalize that human males have more interest in deflowering virgins than females do. In addition, our courtship ritual is perhaps the most complicated in the entire animal kingdom. While practice is unlikely to make a male 'perfect,' we females prefer a man who knows what he's doing when it comes to courtship and sexual relations. Do I want a man who has had a lot of sexual experience? Yes, but only if he's willing to discontinue all other relationships once we've decided to make a go of it. That way I can think of all of his previous experience as practice for the ultimate prize—me!

Buy Me Dinner First

We females are intuitive creatures. We can tell a lot about a potential mate merely by observing him for a few minutes. Physical attributes such as personal hygiene are fairly easy to assess from a brief inspection and provide valuable information to a prospecting female. Other aspects of a male's personality are less obvious but become clear as one spends more time with him. Where does he take me for dinner on our first date? A refined restaurant where he can clearly display his table manners or a drive-thru on our way to the monster trucks?

Whatever a female's preferences may be, she can use her powers of observation and intuition to assess whether a potential suitor is going to measure up. Outside of the actual venue selected for a date, the behaviour of a potential mate during the date says a lot about what kind of partner and father he will be. Does he smack your hand when you attempt to sneak a bite of his tantalizing dinner selection or some of his fries, or does he share willingly? Whether he shares his food can say a lot about how much he is willing to share with you, with the children, and with others. Food sharing among unrelated individuals is not common in the primate world, but instances have been observed among chimpanzees and orangutans.[106]

Female orangutans (*Pongo* species) have come up with a way to assess the temperament of potential mates before making a commitment. When food sharing occurs in these organisms, it is most likely to be a male sharing his food with a sexually receptive female. Why? Male orangutans can be coercive and violent, specifically when it comes to sexual relations. Females must therefore have a way to learn something about the male's disposition before selecting him as a mating partner. In a detailed field-based study of two orangutan populations in Sumatra and Borneo, biologists examined food sharing behaviour between the sexes. Food sharing was defined as the transfer of a food item collected by one individual to another individual out of the first individual's hand, foot, or mouth.[107] The males' reaction to food-taking by a female ranged from tolerance, to aggressively defending the food, to trying to take food from other females, all of which provided information about his disposition. It was concluded that females would take food from a potential mate in order to assess the risk of violence she might face if she paired with him.

A few lines of evidence support the conclusion that females test out potential mates by helping themselves to some of his lunch. First, there was no direct trading of food for sex. The sharing of food and later eating it by both the male and the female was not followed by a sexual encounter between the pair, suggesting that females aren't merely offering up sex for goods. Second, food items shared from males to females were not highly prized, difficult-to-obtain items. The females and males had an equal opportunity to obtain the foodstuffs that they were sharing. Third, lactating mothers did not attempt to take food from males. The only females who actively solicited food from males were sexually active and available, suggesting that food-taking serves a function in the selection of a potential mate.

Primate females may be the only females in the animal kingdom that utilize the food-sharing behaviour of a potential mate as a way to assess his appropriateness. Males, take note: females use their power of intuition to assess whether you might make an appropriate partner, sexual or otherwise. You don't have to share your fries, but if you don't, you run the risk of losing a potential mate. Sure, we can get fries of our own, but we want some of yours. Deal with it.

It Doesn't Pay to Be Jealous

Humans are generally a monogamous species. A large proportion of us form pair bonds that result in sexual monogamy with a single person for the majority of our reproductive lives. However, this does not mean that we practice *social* monogamy; many of us engage in nonsexual relationships with other humans both male and female outside our monogamous pair. Why not? Friends are one of the major joys in life. They provide us with support and camaraderie, which can go a long way to increasing our overall well-being and hence our biological fitness. In addition, friendships are generally free of the sexual and parental commitments that we share with our sexually monogamous partners.

I am lucky to be in a wonderful marriage with a husband who supports my socially polygamous habits. I enjoy spending time with my friends when I can sneak away from my 'duties' as a wife and parent.

However, I have a few girlfriends who are not as lucky. The husband of one friend becomes jealous and pouty when his wife isn't devoting her full attention to him. Could there be a biological explanation for his behaviour? If he were a red-backed salamander (*Plethodon cinereus*), the answer would be a resounding yes!

Red-backed salamanders are ubiquitous to forests in eastern North America, where males and females form long-term (more than three years) monogamous pair bonds and defend their territory from other salamanders. Experiments were designed to assess the reaction of males to their female counterparts when she had been exposed to other males in a nonsexual, socially polygamous encounter.[108]

Salamanders use phenomenal cues to mark their territory and to recognize the territories of others. Females who had been experimentally exposed to pheromonal cues from a male other than their monogamous partner in a socially polygamous scenario were subject to aggressive punishment from their monogamous partners in the form of biting and threat displays. By comparison, it was business as usual for females that were not exposed to any 'foreign' pheromones in the socially monogamous scenario. The males sent a clear message to their female partners: they don't approve of their socializing with other males, whether for sexual purposes or not, and they will take aggressive action in order to reinforce this message.

It is as yet unknown whether this intimidation behaviour strengthens or dissolves the monogamous pair bond in this salamander species. From the male perspective, it makes sense to be aggressive towards a female if it decreases the probability that she will become sexually polygamous, but he has no guarantee that it does. In order to shed some light on the effectiveness of jealous behaviour, biologists undertook a DNA analysis of male and female salamanders in pair bonds and of all juveniles hatched within the territory. This was done on a large scale, testing for relatedness between fathers and offspring in several monogamous pairings.[109]

Lo and behold, the sperm of a male in a socially monogamous relationship was found to sire only 9 per cent to 50 per cent of the offspring of his socially polygamous female partner. In other words, it's unlikely that female red-backed salamanders are just having a coffee with their male friends. They're sexually polygamous as well as socially polygamous, just as their monogamous mate expected. A monogamous husband has good reason to be jealous; unfortunately, his aggression toward his adulterous partner by means of biting and threat displays seems to be too little too late.

Perhaps my girlfriend's husband, and all jealous husbands, should heed this warning: jealous behaviour doesn't reinforce monogamy in the animal kingdom. In the case of the red-backed salamander, it may do just the opposite, which is bad news for a male's biological fitness. For the salamander, it's impossible to say which came first, the sexual polygamy or the aggression. However, in our species we have a multitude of ways to identify threats to monogamy so that they can be dealt with before any sexual polygamous actions occur. To be jealous without any reason for doing so is asking for trouble when there may not be trouble at all.

Bad Boys

The human female is a complicated creature. The selection of an appropriate mate is often an extremely difficult process that can lead to sadness and heartache when we get it wrong. It seems so simple in theory: choose a mate who is most likely to provide you and your offspring with the best possible combination of genes and other resources. Is this what goes through our minds when a 'rebel without a cause' type of suitor drives up on his motorcycle and asks if we want a ride? Do we dutifully say no and proceed to date the boy next door who has a steady income and promises of fidelity?

I'll let you make your own choices on that one, but let's suffice it to say that there are at least a few of us who will opt for Bachelor No. 1 without a second thought. He doesn't offer the resources we should demand of a potential mate, but we choose him all the same. It may be the excitement of the situation or the thought that we might be the one who can settle down the wild child and make him into the suitor of our dreams. Are we alone in our longing for the bad boy when it comes to mate selection? Do any other species opt for a lower investment of resources from a potential mate just because he looks better at first?

Biologists studying mating behaviour in tree crickets (*Oecanthus nigricornis*) have provided us with some reassurance that we are not the only females in the animal kingdom who have a tendency to go for the bad boys in the crowd. Tree crickets engage in some interesting sexual behaviours: a male provides a proteinaceous food gift for the female to enjoy during copulation. Hearing a potential mate's call, a female will investigate and then mount him (assuming he is acceptable to her) to begin the food secretions from glands on his dorsal side. While she's busy with her meal, the male copulates with her, transferring a spermatophore to her spermatheca. She continues to feed while the sperm makes its way from the spermatophore into her storage organ, but when she finishes feeding she reaches back for the spermatophore (or what's left of it), and eats it. The obvious message here is that males who provide a larger food gift will likely achieve more successful fertilization because the female will take longer to ingest it.

Where does the bad boy preference come in? According to biological dogma, females should prefer males that give them the most direct benefits, in this case the largest meal. However, female tree crickets have been observed to prefer larger males despite the fact that they don't give her a proportionately larger gift. In fact, the gifts from these bad boys often are not even large enough to ensure complete sperm transfer.

What's going on here? Why are the females wasting their time with a suitor who may not have any seeds left in his cache? Experimental analysis demonstrates that this is a classic case of females all making the same decision when it comes to their boyfriends (see I'll Have What She's Having). Larger tree cricket males provide a more protein-rich gift than smaller males, so females line up for a chance at it. Unfortunately, with multiple successive matings, males' food gifts become depleted, so little is left for females who are mated last.[110]

Does this in itself make the large male tree cricket a bad boy? Not really—the popular ones just get tapped out. However, the story gets more interesting. Curious about what would happen if the large males perceived a greater mating frequency, biologists manipulated the sex ratio to test how the males might allocate their protein meals in the presence of a larger number of females.[111] It was found that in the presence of more available females, the large male cricket decreased his protein allocation to each female with whom he mated. That is to say, when he thought he could score more often, he promised less to each naïve female. Clearly he's a cricket bad boy.

When the sex ratio was skewed towards more males present, the gifts were larger, indicating that the males could assess their chances of obtaining subsequent copulations and adjust their resource giving accordingly. All this from a lowly tree cricket!

The million-dollar question is this: Why do the female tree crickets continue to select the males with the smaller gifts? Can they somehow sense that they are genetically superior to other potential suitors and take what they can get? Ongoing research on cryptic female choice in these organisms might provide some evidence one way or the other, but until then we should be satisfied that tree cricket females, like human females, are sometimes swept off their feet by the wrong guy.

The Bachelor Pad

What *Homo sapiens* female hasn't been disgusted by the stinky socks, crumb-filled couches, and empty beer bottles that are commonplace in the apartments of the men we date? Did their mothers not teach them how to scrub a toilet or make a bed? Are they unaware that we're afraid to sit down for fear of soiling our newly dry-cleaned dress? One of the things that I appreciated about my husband when we first started dating was that his bachelor pad didn't conform to these scary standards. He had a cool little corner apartment with funky posters on the wall and some interesting décor. It was clean and grownup and a great place to hang out. It reflected his personality, and it attracted me to him. Why is it so rare for human bachelors to create this kind of self-reflective environment? If they only realized how much we females appreciate such things, then perhaps it would become more common. When it comes to wooing females with an appropriate space for mating, males of our species have a lot to learn. Perhaps they ought to look to the animal kingdom for a little advice.

Satin bowerbird (*Ptilonorhynchus violaceus*) males construct elaborate homes out of twigs and other organic material. These structures, appropriately called bowers, are generally composed of two parallel walls that arch toward each other to form an enclosure. Male bowerbirds spend a good deal of time and energy building their bowers and adorning them with objects such as snail shells, berries, glass, and white stones.[112] In addition to natural materials, human-discarded objects like keys, kitchen utensils, and metal tags find their way into the décor of nature's ultimate bachelor pads.

Male bowerbirds have a purpose when it comes to decorating their homes: females have been shown to respond to the design of the bower and the number of ornaments that adorn it.[113] Males therefore work to create a space that will be most enticing to the ladies. Competition for the most alluring bower can be intense; males have been observed stealing rare objects from the bowers of their competitors[114] to use in their own. Among the flashiest and most highly coveted components is a feather from the Australian blue rosella (*Platycercus* species) that moults only a few of these beautiful blue embellishments each season.

So how's a girl to choose a suitable accommodation from an array of fancy bachelor pads? What should she be looking for in a potential mate's bower? It turns out that exceedingly rare or difficult-to-obtain objects don't necessarily make her heart flutter. Field observations of an individual male's mating success based on the number and type of trinkets in his bower showed that rare objects were poor indicators of reproductive success.[115] In other words, boys, forget about boosting elusive treasures from your neighbour on the next tree. Just put some effort into creating a space that's clean and has some nice things to look at. Are all you bachelors out there listening?

One more thing I suppose I should mention: the mating system employed by bowerbirds is described as a nonresource based polygyny.[116] The male doesn't provide any resources directly to the female other than his genes. No postreproductive effort, no parental care, not even a cuddle. The *polygyny* part of the description indicates that males mate with as many females as possible, which makes sense if there is no effort required of him once the deed is done. So what else has a male bowerbird got to do? He might as well put all his energy into creating the most female-friendly space possible in order to maximize the chances of spreading his genetic blueprint across the bowerbird landscape. The smelly socks and crumb-filled couches in the *Homo sapiens* bachelor pads might not be so terrible after all.

He May Be a Wuss but I Love Him Still

Luckily for males of the human species, the great diversity of females out there means that most individuals have a decent chance of finding that special someone. Whether you're a quiet professional or a robust tradesman, there is likely a female out there who will prefer you over all others to be the father of her offspring. This kind of diversity in mate choice is not generally observed elsewhere in the animal kingdom. Females of many species show a distinctive fondness for males of a certain kind: the dominant ones. Such a choice often makes good biological sense. Dominant males are associated with access to abundant resources or territories, and they can give a female the indirect benefit of a strong genetic makeup. That's not to say that subdominant (i.e., wussy) males have nothing to contribute genetically or are never reproductively successful; they just aren't generally the female's first choice. It's a good thing the natural world is so diverse, for there will always be creatures that fall outside the norm, and they too need to reproduce. Some females only have eyes for subdominant males despite the fact that they have only limited material or genetic benefits to offer when compared to dominant males. Why would a female choose a lesser male if Mr. Dominant is banging down her door for a chance to mate?

For females of some species, courtship and copulation are not easy or fun—just ask a female Japanese quail (*Coturnix japonica*). Male quails engage in lengthy courtship rituals that consist of chasing their target female, continuously pecking her head and body, seizing the feathers on the back of her head, dragging her around by the feathers, and repeatedly jumping on her back.[117] There isn't exactly a lineup of willing females waiting for their chance to be courted. A direct correlation has been shown between the level and duration of abusive behaviours and the dominance rank of the male performing them: the more dominant males are more aggressive in courtship.

A female must weigh the costs and benefits of mating with a dominant male vs. a subdominant male, and in the case of the Japanese quail, the latter gets the girl. In experiments designed to test the preference of female quail for either dominant or subdominant males, researchers had a female witness a fight between two males. Once the fight was over and a clearly defined winner and loser had been determined, the female was allowed into the fight arena to select her partner. The females overwhelmingly selected the wussy males who lost the fights.[118] The loser males may not have access to the resources that the winners do, but they are not as aggressive towards their female partners. The females prefer to mate with less aggressive males in order to protect themselves from courtship abuse by dominant males The story is a little different for the Pacific blue-eye (*Pseudomugil signifer*), a small freshwater fish that lives in the streams of eastern Australia. Dominant males of this species don't necessarily win the affections of sexually mature females either, but unlike the female Japanese quail, female blue-eyes aren't subjected to aggressive courtship rituals. These female fish have a different reason to avoid the dominants: they don't make the best fathers.

Courtship behaviours displayed by male blue-eyes communicate information to females about their paternal competence: generally speaking, a better male courtship routine equals a better father figure. Females prefer males who spend longer periods of time on their courtship rituals, and they have been shown to experience greater egg-hatching success when they mate with those males.[119] It just so happens that the males who spend more time on courtship behaviours are those who are physically weaker—the ones that lose fights with the dominants. The wussy males may get beat up on the playground, but their increased effort during courtship gives them the last laugh. For this species, dominance and attractiveness are not synonymous, which is great news if you're a subdominant blue-eye male.

In our species, males come in all kinds of physical forms and from all kinds of genetic backgrounds. There are females out there who prefer macho males who have an aggressive physicality and may intimidate other males, but there are also many females for whom such displays of testosterone are repugnant. The unifying characteristic of females in all species is that we are ultimately searching for mates who will provide us with the most direct (and indirect) benefits at the smallest cost. To put it bluntly, boys, your genetic benefits are too expensive if you're going to abuse us and make us suffer in order to obtain them. The female *Homo sapiens* is much more likely to choose a gentleman when it's time to fulfill her biological destiny.

Cross-dressing

I know several men who like to dress up like women; it's not a phenomenon that is restricted to one kind of sexual orientation or demographic. The truth is that cross-dressing in the human species is widespread. Having grown up in Vancouver, a liberal city, I've seen many a drag queen who kicks my butt when it comes to being the most stunning woman in the room.

Male *Homo sapiens* who dress up as females are definitely not alone in the animal kingdom. Males of several diverse animal taxa like to appear in drag during certain stages of their lives. From insects to fish, birds, and mammals, we see examples of cross-dressing. Several different hypotheses attempt to explain why some males have evolved to practice female mimicry, mostly involving interactions between subordinate males and dominant males. But how could it be advantageous to appear as a female?

Cross-dressing has several possible functions for an inferior male. It can fool a dominant male into permitting a subordinate close contact with females; it can allow lesser males access to high-quality territories they couldn't breach on their own; and it can serve to reduce overall aggression within a group. Any or all of these reasons may apply, depending on the kind of organism doing the cross-dressing and the social structure to which it adheres.

Individual parasitoid wasps (*Lariophagus distinguendus*) undergo their developmental process in stored grain (much to the chagrin of the humans storing the grain). Females oviposit into a single piece, where each individual wasp develops, protected inside its own home and clustered with others. The females in the cluster, while still in their individual grains, produce pheromonal signals that help males find them. This event is crucial, because each female mates only once. A male wasp who detects her signal will wait nearby for her to emerge, in order to inseminate her before any other male gets the chance. (Can you imagine a female of any species mating just as soon as she's hatched?)

This is all well and good for males that emerge early from their own grains, but what about those that emerge later? The late-bloomers need a fair crack at the females as well, so they have evolved a special technique to improve their chances: appearing as females. Male wasps mimic the female pheromonal signal while they are still within their grains, effectively causing other males to sit and wait for someone of the wrong sex.[120] The fact that some early-hatching males have been distracted by the fake female pheromone means that the cross-dressing males have a greater chance of spreading their seed to the next generation. This is cross-dressing in an olfactory rather than a visual sense, and that's the one that matters if you're a parasitic wasp.

Capuchin birds (*Perissocephalus tricolor*) take cross-dressing to the next level. Behavioural observations on this South American rainforest species have demonstrated a great deal of sexual mimicry. Males appear as females, but females also appear as males.[121] In this extreme form of cross-dressing, subordinate males benefit by getting closer to true females, who are normally under the protection of a dominant male.

Copulation in this species takes just a second or two, so a male stands a pretty good chance at copulating if he can make his way to a sexually mature female without being detected. Females benefit from appearing as males because they're not bothered by subordinate males whilst on their quest to copulate with a dominant male. Generally, subdominant males continually chase females, and she can get a break from being pursued by appearing as a male. Finding a partner in this species sounds like a daunting task!

By appearing as a member of the opposite sex, individuals of the species in the examples above have evolved ways to make themselves more successful in the reproductive sense. Can this same logic be applied to sexual mimicry in the human species? Do cross-dressing men increase their chances of offspring production by using their disguise to invade another man's home? It doesn't seem very likely. It seems to me that this is an example of human behaviour without any biological basis at all. Reasons for cross-dressing in our species have little to do with reproductive success or increased survival. It may simply be that the elaborate garments generally worn by the female gender of our species are too lovely for some males to pass up. After all, who doesn't want to feel pretty now and then? I know that I'd certainly prefer a fabulous dress over a boring old suit any day.

I'll Have What She's Having

Members of the human species are notably influenced by what others think. We seek approval from our peers on just about everything: what we wear, what we do, and of course, the mates we choose. Mate choice shouldn't be subject to group approval. After all, if you've found someone with whom you're compatible, why not just snap them up? Females seem to be especially susceptible to the opinions of others when it comes to selecting Mr. Right. Not only do we seek the opinions of our girlfriends for approval of a potential mate, we observe other females in order to determine just what an appropriate mate choice might be. If a certain male is being courted by a female, others sit up and take notice. They assume that the courting female is privy to important information about this potential mate, and suddenly he's the most popular bachelor in town. Popularity can outweigh other characteristics in the selection of mates, despite the fact that it may or may not have anything to do with biological fitness. Are we human females alone in our mixed-up approach to mate selection? Of course not! Females of several animal species are influenced by the mate-choices of their peers, so much so that this behaviour has an official name: *mate-choice copying*.

Female Caribbean guppies (*Poecilia reticulata*) have been shown to engage in mate-choice copying both in the wild and in the laboratory. When females were presented with a choice of male partners in an experimental setting, they showed an overwhelming preference for males that were associated with another female,[122] demonstrating the influence of popularity for mate choice in this species. Female guppies of the sailfin mollie (*Poecilia latipinna*) demonstrate sophistication when it comes to mate-choice copying. Not only are these gals capable of selecting a mate based on his popularity with other females, they also discriminate against males based on the quality of females who associate with him. Laboratory experiments designed to investigate the level of mate-choice copying in this species revealed that an unpreferred male could become a preferred one if he associated with a high-quality female. Further, a male who was once preferred could lose his appeal by associating with a low-quality female[123] (talk about cliques!). So impressionable is the sailfin mollie that she will reverse her own choice of mate based on his subsequent association with other females.

There are several hypotheses for the origin and evolutionary maintenance of mate-choice copying. Younger females have often been seen copying the choices of older females, leading to the logical conclusion that one can gain a reproductive advantage by following the choices of more experienced individuals.[124] In addition, copying the mate choice of others can cut down on mate-searching time and the costs associated with it,[125] such as predation risk and energy loss. This is all well and good if you're a female guppy or sailfin mollie, but what if you're a female *Homo sapiens*?

The critical difference between us and them in this case is our overall life strategy with respect to the mates we covet. Both guppy and sailfin females (and other fish and bird species documented to exhibit mate-choice copying) have reproductive strategies that include little or no parental care investment by the father. Females require only one thing from the male: DNA. Grab it and go! Humans, on the other hand, want much more than that. We want the DNA and the machine that made it; we want the entire package for at least the foreseeable future. Why, then, do we select mates that are popular to other females? If we want the entire package all to ourselves, it makes no sense to copy the choices others are making. I shake my head at the biological anomaly that is the human female. No wonder there are so many single gals out there.

Chapter 5

The Deed

Yes, I'm On the Pill!

Among the most non-biological things that *Homo sapiens* females do as members of the animal kingdom is to block our chances at reproduction for many of our prime child-bearing years. More than 50 million women take the contraceptive pill worldwide. Its physiological effect is to fool our body into thinking it's already pregnant, suppressing the process of ovulation and making pregnancy impossible. Although only females take the pill, males knowingly and enthusiastically take part in sexual relations with these infertile females. The males contribute to this self-inflicted sterility. From a biological standpoint, it's just wrong.

Why do we do this? For most females who are on the contraceptive pill, and for the males copulating with them, the point is to avoid becoming parents before we're ready. The human female is unique in our major time gap between when we're biologically ready to bear children (as soon as we hit puberty we're good to go) and when we become psychologically ready for the long journey through parenthood. There can be a decade or more between these two events, making self-inflicted sterility a reasonable choice. Further, those of us who are ready to embark on the journey of motherhood don't necessarily want to max out our reproductive potential by having as many children as our bodies will allow. Self-inflicted sterility is a desirable option for this reason as well. At times the actions of *Homo sapiens* are unparalleled in their ignorance of biological principles, but this isn't one of those times. We're not the only girls in the animal kingdom to exercise self-inflicted sterility.

The feeding behaviour of two tribes of baboons (*Papio hamadryas anubis*) in Nigeria is a case in point. Some aspects of the females' diet may have implications for how we view the reproduction aspect of the 'survive and reproduce' mantra. Over 1,000 hours of field observations and detailed analyses of fecal hormone content revealed substantial seasonal fluctuations in the progesterone levels of females, attributable to one specific food source: the African black plum (*Vitex doninana*).[126] Baboons are omnivorous, ingesting a wide variety of foodstuffs depending on location and season, but females of both tribes love the African black plum. When it was in season, the black plum was a highly sought-after food source.

What's so special about is this plum? It contains extremely high levels of progestogen, the main hormone that prevents ovulation and the expression of female sexual swellings. When female baboons become fertile, they display distinctive swellings in the anogenital region that serve to attract males for courtship and copulation. The black plum suppresses this process, making it both a physiological contraceptive (since ovulation is suppressed) and a social contraceptive (since it's the swellings that attract the males).

Biologists argue that the African black plum may have other medicinal properties that counterbalance the fact that it causes a seasonal sterility. We generally assume that members of the animal kingdom wouldn't compromise their reproductive output without a good reason, but I wonder if female baboons might be like female humans—they just want a break. Is it out of the question that our primate cousins enjoy the freedom the black plum brings? After all, in plum season they don't have to fend off amorous males. What about the possibility that they appreciate a season free of the physiological costs of child-rearing? It could be elitist of biologists to assume that humans are the only organisms to appreciate that there's more to life than just passing on those blueprints. Baboons are among our closest biological relatives—let's cut them a little slack!

No Eggs? No Problem!

According to our biological mantra, any energy spent finding, courting, or fornicating with a member of the opposite sex is only justified if it's geared toward spreading one's genetic blueprints. Where's the fun in that? *Homo sapiens* males engage in copious amounts of sex without any interest in reproduction. They actively seek out partners who are sexually sterilized so they don't have to worry about inconvenient side effects (i.e., offspring). Biologists generally assume that most other boys in the animal kingdom are much more astute than this when it comes to leaving their share of genes in the pool for subsequent generations. Indiscriminate sex should be a rare occurrence, since it means energy is wasted on dead-end sex instead of being used for other forms of survival (e.g., food gathering or avoiding predators) or reproduction (e.g., courting a viable mate or creating a favourable environment to attract one). But there are others, like the human male, who don't necessarily wait for Ms. Right to come along before attempting sexual relations.

Animal species that live in both sexual and asexual forms present an interesting conundrum when it comes to mate selection. Females are generally the ones who have both sexual and asexual morphs, leaving the male to determine where his sperm will be most usefully spent. However, many males cannot tell the difference between sexually competent females and sterile females. Human males aren't the only ones to discard sperm without regard for the future.

The New Zealand mud snail (*Potamopyrgus antipodarum*) is a lake-dwelling mollusk whose females can be either sexually reproducing (requiring male input for successful embryo production), or asexually reproducing (clonally reproducing without sexual activity). Further, many native populations of this organism are infected with a parasitic trematode that causes castration (sterilization) in females. Hence, males in these populations have several factors acting against their sexual success, leaving them with quite a conundrum when it comes to allocating energy to reproduction. One might imagine that the powers of evolution would have dealt these poor fellas some assistance in the mate-discrimination department, but that doesn't seem to be the case. Mate-choice experiments in which males were given a choice of either sexual vs. asexual females or healthy vs. castrated females revealed that they don't discriminate.[127] Males showed no preference for viable over nonviable females, instead simply attempting copulation with whichever females they could find. In this species, the average copulation event lasts approximately two hours, during which both the male and the female are relatively immobilized, leaving them more susceptible to predation. It's clear that a copulation event represents a fairly large cost to a male if he mates with an asexual or sterilized female. The possibility exists that there may be an even larger cost to a male in terms of time and energy lost if he were to attempt to discriminate between fertile and sterile females, but the researchers surmise that at some level the male mud snails are engaging in behaviour that just doesn't contribute to their biological fitness in any way.[128]

Rotifers are tiny freshwater-dwelling organisms that have two distinct female forms, sexual and asexual. Akin to the mud snail and the human, there are no clear physical differences between the two kinds of females, but those that are sexual must be fertilized at a very early age. They are no longer fertile after 9 to 20 hours of life.[129] Male rotifers show a distinct preference for fertilizing very young females two to three hours old, which slightly improves the likelihood of fertilizing a sexual female, but they don't specifically discriminate between sexual and asexual individuals.[130] Why don't the males select females who can propagate their genetic lineages?

They have a short lifespan—about 48 hours—and a large enough supply of sperm so as not to become completely tapped out during this short time. It takes about 13 copulations for them to be spent. This drastically reduces the need to discern between sexual and asexual females. If they had less sperm to work with, they might feel more selection pressure to find the right girl rather than any girl.

As these examples show, if a male cannot distinguish between fertile and sterile females, several of his sexual conquests may be in vain. This could mean big trouble if you're a rotifer or a mud snail. Reproduction is as important as survival to the individual, and if the chances to reproduce are impaired, biological fitness is automatically lowered. So where does that leave *Homo sapiens*? Unlike many of our cousins in the animal kingdom who are mostly focused on reproduction, a great many humans prioritize *minimizing* biological fitness with regard to reproduction. Human males, unlike their snail and rotifer counterparts, actually seek out sterility in a potential partner, and for good reason. Can you imagine all of your own conquests resulting in offspring? You might have greater biological fitness than all your friends, but to *Homo sapiens* this situation is horrifying. Imagine how many weekend soccer games you'd be scrambling to attend!

The Chastity Belt

It's a tough world out there. For the average *Homo sapiens*, having security in your relationship is critical to feeling successful. If someone loves you enough to be faithful to you and to forsake the advances of all others, it reinforces your self-confidence and makes you happier and more productive human. In the real world, however, not all pairs adhere strictly to the desired faithfulness. What happens if you're feeling less than secure about your partner's fidelity?

In the fifteenth century, humans invented a nifty contraption that assured insecure males that their partners wouldn't accept genetic donations in their absence: the chastity belt. Made out of tough material, usually a combination of steel and leather, the belt covered the female's private areas and prevented any kind of sexual interaction. It seems a rather drastic measure, but as I said, it's a tough world out there. It turns out that the chastity belt method of preventing one's mate from copulating with others isn't limited to our species.

A *mating plug* is a structure that blocks the female genital reproductive tract. Males of many species, including insects, crustaceans, reptiles and even mammals, use such plugs to prevent further mating by a female once they have deposited their DNA into her reproductive tract. In species where there is intense sperm competition, biologists have described some sophisticated mating plugs. The spermatophore of ground beetles (*Oedothorax retusus*) contains sticky substances that allow it to act as a mating plug by physically blocking the female's reproductive opening. In addition to physical blockage, the seminal fluid of ground beetles induces refractory behaviour in mated females: it makes her avoid copulating with other males.[131]

Dwarf spider (*Leptocarabus procerulus*) males utilize glandular secretions that harden once they are deposited inside a female. There is an extremely high level of sperm competition in dwarf spiders, as the sperm from a single mating event can remain fertile in a female's sperm storage organ for several months. The dwarf spiders' secretion plugs are highly effective at preventing further mating by females: an experimental study showed that a large plug produced by an uninterrupted copulation prevented further mating by the female 93 per cent of the time.[132]

If you think that the chastity belt strategies of male ground beetles and dwarf spiders are shocking, read on. In arachnid species where sexual cannibalism is demonstrated (these are species where the female kills and ingests the male after copulation), males go even further to ensure paternity. This makes sense. If he's going to die anyway, he might as well do whatever it takes to make sure that his DNA is the prize-winning seed for the next generation, even if this means breaking off his copulatory device inside the female genital opening.

Males of the orb-web spider *Argiope lobata* and the white widow spider *Latrodectus pallidus* have been shown to do just that.[133] Plugging up the female with their copulatory apparatus has the effect of blocking the opening after the sperm transfer, and it has been experimentally shown that such does reduce the paternity share of males attempting to mate with a female once a plug is in place. Further investigation of sexual behaviour in this species demonstrated that males who were cannibalized on their first copulation attempt had a much higher probability of damaging their pedipalps (sexual sperm-transfer appendages) than males who escaped (74 per cent vs. 15 per cent).[134] Since this self-induced damage has negative consequences for future reproductive bouts (no kidding!), it makes the most sense biologically for a male to do it if his chances of securing paternity are increased.

Although males of several species have developed foolproof ways to ensure the fidelity of their sexual partners in order to guarantee paternity of the offspring and maximize fitness, I am much relieved that in our species such methods are almost unheard-of. The chastity belt is a thing of the past, having been replaced by trust in one's partner and vows of monogamy. I think that both male and female humans would agree that these methods are a much better arrangement than losing your copulatory organ or having your opening plugged up indefinitely.

Not Tonight, Honey, I Have a Headache

To put it plainly, we human females don't always feel like having sex. Sorry, males, but sometimes we'd rather have a back rub or just be left alone to rest. Sometimes we want to use our energy for a different purpose. Most of the sexual encounters that take place between monogamous human couples are not for the purpose of procreation, so from a biological perspective the lack of interest displayed by females makes sense. However, if the only sex had by humans was for baby-making, there would be a lot of sexually frustrated individuals out there, since the average gestation time for a human embryo is nine months. Most human females don't bear as many offspring as our physiology allows; most of us pull the child-rearing plug after one or two or three, despite being fertile for more than thirty years. This creates a 'need' for members of our species to engage in sexual behaviour more often than is necessary to create offspring. There are benefits to engaging in such behaviour, both physiologically (to avoid the frustration that results from toxic sperm buildup in males) and psychologically (for pleasure and to strengthen the pair bond, avoiding the possibility of extra-pair sex). So yes, there are reasons to have sex for nonprocreative purposes, and while the advances of our male partners are welcomed and enjoyed for the most part, I reiterate: sometimes we just don't feel like it.

Female water striders (*Gerris* species) are another animal species whose females don't always welcome the sexual advances of their males. Actually, this is a gross overstatement. These gals almost always have a headache, and for good reason. Mating behaviour in several species of water striders represents a dramatic conflict between the sexes. The female has the capacity to store a large amount of sperm for up to ten days, meaning that with relatively few copulations she can acquire all the sperm she needs for her entire reproductive lifespan.[135] As you might imagine, this presents a problem for males who have not yet had a chance to copulate with her.

Further, there are advantages to being the last male to inseminate a female before she lays her eggs. It has been demonstrated that up to 80 per cent of a female's eggs are fertilized by the last sperm she receives,[136] and there are many males competing to be the last to contribute to her sperm bank, regardless of whether the bank requires another deposit. This is a bleak situation for females who have already had their fill. To add insult to injury, unnecessary sexual activity comes at substantial cost to a female. Females carry the males around on their backs during copulation, which is a strain both directly in terms of the energy required to carry him and indirectly in the lost energy that could have been spent accomplishing other tasks. In addition, the monkey on her back impairs her mobility on the water, causing her speed to decrease and her visibility to increase, both of which make her more susceptible to predation. For all these reasons, females actively deter superfluous mating attempts by males, either by fleeing from their advances or by attempting to make copulation more difficult.

The dichotomy in the requirements of male vs. female water striders has resulted in an evolutionary arms race of behaviours and structures that help members of both sexes to remain reproductively successful. Males have evolved elaborate grasping structures with which to entrap a female until the deed is done, but females have evolved complex antigrasping structures that work to deter the male or to dislodge him if he successfully mounts her. The level of armament between males and females of the same species has been shown to be closely correlated,[137] meaning that the evolution of these structures in both males and females correlates to the behaviour and success of the sexual partner.

Luckily, the sexual interaction between males and females of most animals, including humans, doesn't involve such a dramatic conflict. For the most part, needs are met without elaborate behaviours and mechanisms with which to avoid having sex. In our species, paternity is generally agreed upon before the fact, obviating the need for males to engage in coercive attacks in order to pass their blueprints to future generations. Next time you find yourself hoping your partner would direct his sexual energy elsewhere, be thankful that you aren't a water strider.

Artificial Insemination

Upon first consideration it might seem unnatural for a human female to turn to a sperm bank in order to propagate her genetic lineage. Yes, natural sex is removed from the equation, but when it comes to the selection of a donor, she can be choosy with respect to physical and behavioural characteristics like race, physical health, and even the IQ of the male with the winning seed. In a perfect world, we would all define the most important characteristics for our mates, find mates with those characteristics, and procreate in order to obtain offspring with the characteristics we prefer. However, reality in the natural world is harsh whether you're *Homo sapiens* or some other animal, and sometimes things just don't work out optimally.

In organisms where multiple males compete and copulate with a single female (a polyandrous sexual system), females are often coerced into sexual activity with males they wouldn't otherwise choose. What's a female to do if undesirable sperm happens to find its way into her reproductive tract?

Cryptic female choice (CFC) refers to the power of the female to bias sperm use towards that of preferred males, despite the availability of sperm from other (suboptimal) males. Females in several species have evolved ways to allow for the sperm of certain males to fertilize the precious eggs (not entirely unlike selecting seed from a catalogue in a fertility clinic).

Female freshwater guppies (*Poecilia reticulata*) overwhelmingly prefer to mate with males that have bright body coloration, specifically with large orange spots.[138] Do they have the ability to swing the insemination odds in favour of a good-looking suitor? It appears that they do. In laboratory experiments, female guppies were given a choice to mate with an intermediately coloured male in two situations: when he was the more attractive candidate paired with a dull-coloured individual, and when he was the less attractive candidate paired with a very brightly coloured individual. In both cases the only male who had access to the female was the intermediately coloured one—the comparative individuals were visible but not accessible to the female. The results were clear: the intermediately coloured males inseminated 68 per cent more sperm into females when they were perceived as the more attractive candidate.[139] The mechanism for this is not known, but there is no question that the female exercised some control over the number of sperm successfully transferred to her reproductive tract after a copulation event. If the male was attractive, she kept more of his sperm.

Another example of females manipulating the insemination success of various types of sperm comes from the feral fowl *Gallus gallus domesticus* (wild chickens). These organisms have a complex social system, with males in an intricate hierarchy of social dominance. Females prefer to copulate with dominant males and not with subordinate ones, but the underdogs still copulate, often violently coercing the female in order to do so. Fortunately, the females appear to get the last laugh: analysis of the fertilization success of dominant vs. subordinate males showed that females eject the ejaculates of the lesser males after copulation.[140] So although a subordinate male can use his strength to force copulation upon an unwilling female, his chances of paternity are limited, because she can discard his donation in favour of one that she actively seeks out.

In the natural world, there are many examples of females biasing paternity in favour of specific male phenotypes or social ranks, much like a human female in a sperm bank selecting the seed of a successful entrepreneur over that of an unemployed couch surfer. However, the major difference is that in the natural world females can undertake such selection without the intervention of fertility procedures devised by humans.

Even the lowly female chicken, who has proven to be more than just the dumb bird we eat for dinner, displays a level of sophistication that seems unattainable for *Homo sapiens*. In species where coercion is commonplace (and I would argue that ours is one of them), it is extremely advantageous for females to employ mechanisms to avoid having offspring fathered by undesirable sperm. If that means making a well-informed decision after perusing a brochure from a sperm bank instead of making a hasty choice after a few drinks at a club, I'll vote for the former.

Ladies of the Night

Homo Sapiens engages in sexual activities far more frequently than is required for procreation. In fact, many humans have sex for a living, engaging in copulation with others in exchange for some form of payment. Biologically speaking, what is prostitution? It's the trading of sex for goods. In humans, the goods come in the form of cash, food, or other material possessions provided by the male in exchange for sex provided by the female. There are also organisms in the animal kingdom that utilize the power of prostitution obtain goods other than those of the genetic variety. There can be complicated reasons for engaging in copulation outside of the goal of passing on one's genetic blueprints, including the acquisition of goods that are not readily available otherwise.

Many arthropod species have been shown to trade merchandise for sex. They employ a polyandrous sexual strategy (one female mating with several males) described above for several invertebrate species. Females often can obtain all the sperm they need in a reproductive cycle from one sexual encounter, but multiple copulations are normal. Additional copulations can have high costs to females for several reasons, including injury (due to rough or forceful copulation), predation risk, disease, male harassment, and reduced foraging opportunities[141] (see Not Tonight, Honey, I Have a Headache). Then why do females of some species actively seek out superfluous copulations?

Indirect benefits from having more sexual partners may include an increase in the genetic variation of offspring and increased sperm competition in her sperm storage organ (see Artificial Insemination) that confer genetic benefits on her offspring. On the other hand, direct benefits to females for engaging in repeated sexual encounters can take the form of services provided by the male, such as parental care, or the delivery of goods, such as nutrition. If sex takes place for the gain of material benefits as opposed to the gain of genetic material to create offspring, can we call this arthropod prostitution?

Mormon cricket (*Anabrus simplex*) males provide a nutritious meal to their female partners during copulation, which offers direct fitness benefits outside of the sperm provided by the sexual act itself. Studies have shown that in times of food shortage, females actively compete for access to males and thus to the food benefits that come along with doing the deed.[142] When food is plentiful, such competition doesn't take place.

Similarly, in the bruchid weevil (*Bruchidius dorsalis*), males donate nutrition to the female via the seminal fluid. Seminal donations from the male can constitute up to 7 per cent of his entire body weight! Females mate more often when food is scarce or of low quality,[143] indicating that there is an increased need for food (not sperm) donations during such times. If males are well-fed and therefore provide a higher nutritional value in their seminal fluid, females are less likely to mate often, because their nutritional needs can be met by one well-fed mate.

Female seed beetles (*Callosobruchus maculatus*) will acquire goods outside the act of copulation if they are readily available, but it has been shown experimentally that if needed commodities are not easily obtainable, then sex is the next best option.[144] In an experimental manipulation examining the feeding and mating behaviours of seed beetle females who were provisioned with food and water, just water, or nothing at all, it was found that the hydration benefits provided by the ejaculate were what kept the females coming back for more (this must have been a blow to the male seed beetle ego, but I digress). These organisms generally live in arid environments where water is scarce, and females were shown to mate more often when excess water wasn't provided. Mating frequency was lowest when water was supplemented for the females. In this species, there are direct costs involved for females who engage in repeated copulations (in the form of increased male harassment and energy loss), but when important resources such as water are scarce, the costs are tempered by the benefits of extra water.

Comparisons to prostitution practised by the human species can be made at a basic level. In tough times when resources are scarce, female arthropods undertake superfluous copulations in order to acquire material provisions. If resources are abundant and repeated copulations are not necessary to obtain needed commodities, she spends her energy elsewhere. Did prostitution start with the arthropod? From a female perspective, this seems true to the human rationale behind prostitution, but arthropods cannot utilize the term due to differences from the male perspective. Male arthropods, from weevils to beetles and crickets, have one common goal: to fertilize as many eggs as possible with their own sperm. They participate in repeated copulations with females who may be only interested in the gifts they offer so that their sperm has a chance to sire the next generation. Human males who actively solicit sex in exchange for material goods—cash in the case of prostitution—are doing the opposite. In our species, prostitution is for recreational purposes only. Sex in exchange for genetic contributions is generally free of financial transaction.

Monogamy with a Twist!

My husband is a real guy's guy when it comes to his appearance. He's never set foot in a hair salon (he prefers the $8 haircuts he can get without an appointment). There is no talk of hair products or colouration, designer clothing, or professional intervention when it comes to any human male-practised grooming ritual. In my opinion, that's the way it should be. If he started fussing too much with his appearance, I would immediately become suspicious. Why go to all the trouble when we're already married? I would suspect some other reason, such as another female, for any additional grooming or adornment.

A high degree of sexual ornamentation is most often associated with sexually promiscuous or polygamous species in the natural world. Males who are highly adorned provide a signal to prospective females that they're healthy and have high-quality sperm. If you aren't going to settle down with one partner, you'd better always look good in order to keep attracting mates. So how does monogamy fit into the adornment scenario?

Once monogamous partnerships have formed, for a mating season or for life, is it really necessary to maintain elaborate ornamentation? Don't the rules of biology dictate that energy is better spent on collecting resources or contributing to the well-being of offspring? Mother Nature says no. Several socially monogamous species maintain a high degree of sexual ornamentation.

Charles Darwin proposed two hypotheses for sexual ornamentation in monogamous species: if males regularly outnumber females, there should be intense competition between males for their affections, and if females vary greatly in important qualities such as health or fecundity, then males will compete with each other for the opportunity to mate with the high-quality females.[145] Such hypotheses explain the initial contact with a potential mate, but what about after the fact? Why maintain a costly sexual ornament if your pair-bond has already been established? Every female's worst suspicions are confirmed: biologists have documented that in several monogamous species extra-pair mating takes place and can substantially increase the reproductive success (and biological fitness) of the males who participate in it.

The mountain bluebird is a socially monogamous species in which males have a bright blue ultraviolet (UV) coloration of the feathers on their dorsal side and rump. Females are not similarly adorned, having dull blue or grey feathers. Although the males are the ones with the sexual ornamentation, not all ornamentation is equal. Some males are brightly coloured and others are intermediately coloured and presumably not as attractive to females.

The reproductive success of the males with the brightest colouration is higher than that of their duller counterparts,[146] and not just with their monogamous partner. Genetic analysis has demonstrated that the brightly coloured males get more action both at home and elsewhere, siring clutches of eggs in their own nests with their own partners (within-pair mating) and also in the nests of other males (extra-pair mating). So what's a female to do? If her highly adorned, high-quality mate is out there siring clutches with other females, should she just sit around and wait for him to come back? I think not!

A male has been cuckolded if another male invades his territory and mates with his female partner. This is extremely bad news for species of birds whose males provide parental care, because providing care to an offspring not biologically related is a costly endeavour. Males therefore undertake actions that aim to suppress or eliminate the possibility that a more attractive suitor will swoop in and steal his paternal rights. The most common cuckolding avoidance behaviour is mate guarding, where a male remains close to his female partner and defends her against any males that stop in for a quickie.

Like the mountain bluebirds, male grosbeaks (*Passerina caerulea*) are brightly adorned with UV colouration of their feathers, whereas females are a dull brown colour. But not all males are created equal, and as with the bluebirds, the incidence of extra-pair copulations is greater in males who are brightly adorned than those who are dull. One would therefore hypothesize that dull males should spend more time guarding their mates against the advances of their highly adorned counterparts. However, dull-coloured males don't guard their mates to a greater degree than brightly coloured ones.[147] Instead, mate guarding is directly related to the attractiveness of neighbours. If nearby grosbeaks are brightly coloured, males are more likely to guard their mates, regardless of their own colouration. Males can assess the risk of their partner engaging in extra-pair relations based on the number of brightly coloured males in the immediate area and manage their own guarding behaviour (and indirectly their own extra-pair mating behaviour) accordingly.

So ladies of the mountain bluebird and grosbeak species, the moral of the story is this: if you're in a monogamous relationship with a highly adorned male, watch out. The chances of him getting busy with another female, or attempting to do so, are high. However, all is not lost. An enterprising female can use the time when her partner is away to sneak in a few extra copulations herself, provided there are attractive candidates nearby that her partner hasn't noticed.

Does this advice pertain to monogamous relationships in *Homo sapiens*? Well, it could. If your mate starts investing a lot of energy into physically adorning himself and spending a lot of time away from the nest, you might decide to take a stroll around the neighbourhood to see what else is available.

Sneak It in When You Can…

Not long ago I was out with some girlfriends at a local pub. Shortly after we arrived, a waitress approached our table with a tray full of tequila shooters 'from the gentleman in the corner.' It's hard to say what I found more amusing: the proud bachelor sitting a short distance away or the sight of ten 30-something women licking salt off their wrists and giggling as they did tequila shooters for the first time in a decade. The tequila came and went, and nothing else changed, but this event got me thinking about the biological strategy behind the maneuver. What would possess a male to do such a thing? He must feel that what he has to offer a potential mate is of low value or he wouldn't resort to buying ten shots of tequila in the hopes that some female's judgment might be altered enough to allow him a chance at copulation. If he had more to offer, perhaps he wouldn't feel the need to sneak into someone's bedroom this way.

Sneaking a mate is a common occurrence in the animal kingdom. Males who are unable to offer much in terms of resources to woo a female often resort to sneaking in a copulation here and there. Take the coho salmon (*Oncorhynchus kisutch*) as an example. Males of this species spend a year in a freshwater stream as juveniles before heading out to the ocean. They then migrate out to the ocean for a period of either 6 months or 18 months before returning to their natal stream to spawn. Needless to say, the males who spend only 6 months in the ocean end up considerably smaller in size and stature than those who spend 18 months. These two male morphologies are commonly referred to as jacks (the smaller) and hooknoses (the larger).[148] Male hooknoses can be very aggressive and might attack or even kill jacks if they hover too close to a spawning female. How is a small jack supposed to compete with a large, aggressive hooknose for a chance to fertilize a female's clutch of eggs? By sneaking, of course! In habitats that have lots of debris and other refuges for the smaller males to hide in, jacks can be reproductively successful by waiting for the right moment to dart out and fertilize some eggs before anyone notices.

One species that takes sneaking behaviour to an extreme is the Galapagos iguana (*Amblyrhynchus cristatus*). Males of this species have three possible strategies for mating: holding territories, ranging as satellites, or sneaking.[149] These strategies are largely dependent on the age of the male: the smallest (youngest) males cannot effectively fight against their older, larger counterparts for either territories or access to mates, but they have a distinct advantage for being sneaky in that they look and behave like females. They can inhabit the territories of large aggressive males without being noticed as potential competitors and sneak in a copulation or two when the big males aren't around. The sneaking strategy lasts only as long as the males remain very small. Once they grow a little larger, stray males become discernable from the females and are chased out of the territory by the dominant male. They then adapt a satellite strategy, remaining on the prowl and attempting to grab females for sex while they're en route to foraging sites. On the other hand, the largest and most aggressive iguana males hold territories, defend them, and have no trouble courting (and fathering the offspring of) the females who live within their territory. In other words, these large territory-holding males need not be sneaky, since they've got the goods and the females know it.

I wonder about the existence of multiple reproductive phenotypes in the human male. Some of them have females swooning at their feet and need not resort to sneakiness, while others less fortunate must resort to alternative strategies. Were the tequila shooters merely a way for the bachelor in the bar to hide his inadequacies? Should I be disgusted by this sneaky behaviour, or should I appreciate that his attempt to sneak a mate is a biological strategy based on his inability to court a female the way the big guys do?

Ladies, here's my advice: the next time a random guy buys you a drink in a bar, run, don't walk, to the nearest exit. He's attempting to sneak a copulation from you, which may be indicative of a greater biological inadequacy. Do yourself a favour and wait for the male who has resources to offer you and your potential offspring. Or at least wait for one who will buy you dinner first.

Party of One

We humans are really fascinated with our own genitalia, aren't we? From a very early age, humans are well aware of, and curious about, the special properties that exist in the machinery of our nether regions. Perhaps it's an unconscious appreciation for the parts of our anatomy that will ultimately catapult our genetic code into future generations, but whether we admit to it or not, everyone engages in solo sex.

Masturbation, we're told by health professionals, is a natural and necessary process that is enjoyed by males and females, young and old. Although at first glance one might think that masturbation in males is wasteful because genetic blueprints are being released without a potential egg-target, there may be some biological use for this behavior after all. Sperm expulsion by males is a common occurrence in several vertebrate and invertebrate species, and in some cases it's a strategic maneuver that results in greater reproductive success.

It has been observed in house crickets (*Acheta domesticus*) that males routinely expel spermatophores at times other than during sex. Spermatophores are packages of sperm that are transferred to the female sperm storage organ, the spermatheca, during copulation in many invertebrate species. However, not all ejaculates are equal. Female house crickets are more likely to store *young* sperm in their spermathecas.[150] (A polyandrous sexual system in this species means that multiple males can mate with a female in order to fill up her sperm storage organ.) Young sperm can be superior to older sperm for a number of reasons, including a greater fertilization ability and a higher survival rate in resultant embryos. Due to the fact that younger sperm is dominant in the spermathecas of female crickets, a male's chances at reproductive success are greatly improved if he provides a relatively new package. How does he preferentially deposit young sperm? By discarding the old sperm, of course. Autonomous sperm ejection (cricket masturbation) is the mechanism by which a male cricket gives himself a younger sperm content and increases his chances of successfully fertilizing his partner's eggs.

Not all masturbatory emissions are so carelessly discarded. Male marine iguanas (*Amblyrhynchus cristatus*) in the Galapagos archipelago have developed a strategy for improving their reproductive success via storage of their own ejaculates. As noted earlier (see Sneak It in When You Can), there are large size differences between sexually mature marine iguana males. The largest males are the most aggressive, and they hold territories in which the female iguanas nest and feed. Smaller males must somehow sneak into the territories and copulate with females when the larger males are otherwise occupied. Since female iguanas only copulate once per season, smaller males have developed elaborate strategies to get the most bang for their buck (yes, pun intended) during these rare opportunities. It takes approximately 2.8 to 3.1 minutes of intercourse before a male iguana initiates ejaculation.[151] For the large dominant males, this time interval is easily achieved, and such males have a high rate of successful copulations (95 per cent). The story is not as positive for the smaller males, who generally realize a much lower copulatory success rate and are forcefully separated from their female partners by territory-holding males before ejaculation 29 per cent of the time. However, the crafty underdogs are able to increase their fertilization success by masturbating before having sex and storing the prepared viable ejaculate in small pouches near the penis. Storing this extra sperm that can be transferred to the female almost immediately upon mounting her increases the overall fertilization success of the small male iguanas by a respectable 41 per cent. A stored ejaculate strategy may be common in species having a size-dominance hierarchy and sexual interference. Where there's a will, there's a way.

So why do humans masturbate? The aged sperm scenario seems like a good reason in some cases; controlling for sperm age is a standard procedure in human in vitro fertilizations.[152] However, this doesn't explain the absurdly high frequency of masturbation in males, nor its existence at all in females, also observed in other primate species including bonobos and macaques. Although the clitoris is found in all female mammals, scientists have yet to come up with a suitable biological explanation for it, or for the existence of the female orgasm.[153] The long and short of it may be that it's simply recreational. Let's call it a day on the science and just enjoy its nonbiological value.

A Little Boy-on-Boy Action

Homosexuality presents biologists with an interesting conundrum. This sexual behaviour doesn't have even a remote chance of producing offspring to carry on one's genetic lineage, but it is widespread from insects to mammals and everywhere in between.[154] Why does homosexuality exist? Its sheer ubiquity suggests that there may be many reasons. Some of the more common hypotheses for the evolutionary maintenance of homosexuality include its use in establishing social dominance, so that if a male displays power (in this case sexual dominance) over other males, it might lead to future reproductive success with females. Alternatively, homosexual encounters may confer advantages for heterosexual reproduction in the form of practice before a suitable female partner is found, or it might serve to dispose of aging sperm in order to keep ejaculates in top condition for that suitable female partner (see Party of One). Whatever the reasons may be, there is no denying the fact that animals are not exclusively interested in members of the opposite sex.

Homosexual behaviour is extremely common in the male flour beetle (*Tribolium castaneum*), but biologists have found no evidence either for dominant males obtaining a greater reproductive success over submissive ones (the first hypothesis) or for the greater reproductive success of well-practised homosexual males (the second).[155] So why do these guys have so much sex with each other?

In a small percentage of cases, males who had received an ejaculation from another male and then subsequently mated with a female actually passed on some of their homosexual partner's sperm to her, Yes, you read that correctly. A male who had previously mated with another male (and *not* with a female) maintained a small chance of siring the progeny of the next female mated to his homosexual partner. This is an interesting side effect of homosexuality, but the incidence of third party fatherhood was low, leaving scientists to continue speculating about the sexuality of the male flour beetle.

In some cases homosexuality is more situational than ubiquitous, as seen in the flour beetles. Bearded vultures (*Gypaetus barbatus*) generally form long-lasting heterosexual breeding pairs (alpha male and alpha female) that inhabit a specific territory. However, when resources are scarce, such as during a shortage of suitable territories, an additional (beta) male might join the monogamous pair on their territory because he cannot find or maintain his own. Such uninvited company is not welcomed by the alpha male since he may be unable to prevent the beta male from attempting to copulate with his girl while keeps up with the responsibilities that go with maintaining his territory. The alpha male must therefore strike a balance between carrying out his responsibilities and engaging in aggressive behaviour with the beta male. These polyandrous trios (alpha and beta male along with a female) may exist with a great deal of conflict between the two competing males, but it has been found that in order to minimize this conflict the males engage in acts of homosexual copulation.[156] Such acts are thought to serve as regulators of social tension in these situations, enabling a greater level of cohesion within the trio. If you can't beat him, have sex with him!

In contrast to the long-term heterosexual pairings of the bearded vultures, Laysan albatross (*Phoebastria immutabilis*) females have been shown to form long-term homosexual pair bonds.[157] Approximately 30 per cent of the nests of a study population in Hawaii were found to contain female-female duos; the other 70 per cent contained heterosexual pairs. The lesbian couples engage in both mate guarding and mutual preening behaviours. The pairings can be extremely long lasting: a pair of females on the island of Kauai has been together for 19 years! Males are, of course, a necessary part of the reproductive equation for the albatross females, who lay one egg during each breeding season. Males are involved in egg fertilization, but once their genetic contribution has been made they don't stick around.

As these examples attest, there is so much diversity in the animal kingdom with respect to homosexuality that it is virtually impossible to classify all homosexual behaviours. The human species exhibits homosexual behaviours like those of the flour beetle (boy-on-boy action), the bearded vulture (menage à trois and bisexuality), and the albatross (lesbian couples seeking outside intervention in order to have offspring). Whatever the reasons for homosexual behaviour in any species, it must be kept in mind that generally such encounters do not make up the entirety of the sexual repertoire. Most animals that exhibit homosexual behaviours exhibit heterosexual behaviours as well.

The major difference between 'us' and 'them' when it comes to homosexuality gets us back to the drawing board: unlike most species, many homosexual humans are content not to engage in sexual behaviour for the sake of genetic propagation. There is nothing wrong with homosexuality; its ubiquity in the animal kingdom attests to that. However, what *is* wrong, according to the basic rules of biology for any human, homosexual or otherwise, is to forgo your chance at reproducing when you have the physiological ability to do so.

Chapter 6

The Consequences

The World Is Nice to a Pregnant Lady

Having been pregnant three times in my life, I have grown used to both the ups and downs that come with having a basketball attached to your midsection. It's difficult mobility-wise, sleeping is nearly out of the question, and the need to urinate is constant. However, I've found that the reception I get from the general public, members of my species with whom I do not have a close social bond, is just plain nice. Women let you pass them in line for the public restrooms. People of all ages and sexes hold doors for you, smile at you, and even offer to carry things for you. It's as though there's a general recognition that a pregnant lady is doing her part for the human race and she should be rightly rewarded for the valiant effort she's making—or maybe it's just that people recognize that you're physically exhausted and need a little help. In any case, I find this display of human nature refreshing in a world where much of the time people just forget to be nice. In fact, I greatly appreciate it, because, let's face it, when I'm pregnant, I'm not in a position to wait all that long to go to the bathroom, I feel fat, and I am very, very tired. Late stages of gestation carry high physiological costs, and this is true not only for members of our species but for most others as well. Do other animals treat their grossly pregnant females to similarly altered (nice) behaviour?

For most species, the late stages of pregnancy are not as visually obvious as they are for humans. However, it has been observed in several lizard species that a distinctive colouration reflects a female's stage in the reproductive cycle.[158] When burdened with a heavy load of fertilized eggs (i.e., pregnant, which in egg-laying organisms is termed *gravid*), females develop a bright red colouration on their throats. This colouration signals to her conspecifics that she is in a delicate condition and must conserve her energy for this purpose. Indeed, these lady lizards experience a greater rate of predation from creatures like snakes and birds because of both the bright visual cue and the fact that they are unable to move as fast (a no-brainer to any human female who has tried to run during her last trimester). It is therefore advantageous for the gravid female to be able to communicate to her conspecifics that she is not to be disturbed. It seems that the throat colouration in lizards serves this purpose.

Males of the tropical lizard *Microlophus occipitalis* are less likely to engage in courtship with red-throated females, whether the red colouration is natural (as documented through field observations) or painted on (as documented during experimentation).[159] The bright red colouration provides a clear message to would-be suitors: *Currently Unavailable!* Although they aren't exactly rushing to help her with grocery bags, the males cool their courtship rituals and leave her alone. Without the need to fend off the advances of her male counterparts, the gravid females experience reduced predation risk and reduced energy expenditure, both of which are major advantages for mother and offspring.

Like the lizards, females of several primate species undergo physical changes in accordance with sexual cycles, exhibiting a bright red colouration on the hormone-sensitive areas of the face and anogenital region. In the third trimester of a 5.5-month gestation, female Rhesus macaques (*Macaca mulatta*) display the deep red signals of pregnancy, and they are treated differently by other members of their social group. Lip-smacking behaviour, which is generally associated with aggressive avoidance and appeasement in macaques,[160] is observed by both males and females in response to a pregnant female, indicating that the bright colouration can serve as a warning against aggressive encounters (don't pick a fight with me, I'm pregnant!). In addition, males engage in higher levels of 'self-directed behaviour' when in the presence of pregnant females, indicating that the colouration also serves as a warning against the copulatory advances of males.[161]

There is no question that behaviour in both lizards and macaques in response to females in late stages of gestation carries undertones similar to those in humans. Helping the burdened female so that she can conserve her energy for her ultimate biological purpose is the right thing to do, even if she is no genetic relation. The bottom line: Give the preggo a break; she needs it.

The Wet Nurse

Any woman who has breastfed a child understands the effort involved in doing so. One of the indescribable 'joys' of being mammals is that we are blessed with breasts for the purpose of feeding our newborn offspring. (Although adult human males seem to be fascinated with them as well, breasts have no biological purpose outside of this feeding role.) The costs of breastfeeding are high, not only in terms of the energy required for milk production, but in terms of the potential for pain (e.g., chafing, blocked ducts, infection) and the loss of time that we could otherwise spend on other tasks. It's no wonder that some of us count the days until we can start our babies on solid foods. In some cases the physiological costs are too high and a mother might not have enough milk for her offspring.

While humans have developed some convenient ways to deal with this situation (our accomplishments in science and technology afford us the freedom to feed our children from a bottle), such an option isn't available to the rest of the animal kingdom. Mothers must ensure that an adequate source of nutrition is found in order for the offspring to survive, but at times offspring may be required to take matters into their own hands. A common strategy for an infant to gain a little extra nutrition is to allonurse, or to obtain milk from a source other than its biological mother.

Among lactating mammals, allonursing has been observed in 68 species. It tends to be more common in species that are communal breeders (e.g., rodents) or that give birth in common areas (e.g., seals, sea lions). The burning question is this: why would a mother allonurse an infant that is not her own? As I've discussed, there are major costs involved with nursing your offspring, most importantly in terms of energy loss. Empirical evidence is consistent with three different hypotheses for allonursing:[162] misguided parental behaviour (an inexperienced mother may not realize that the offspring on her boob isn't hers); inclusive fitness benefits of nursing-related offspring; and evacuation of excess milk after feeding one's own offspring.

Based on a long-term observational study of steller sea lions (*Eumetopias jubatus*) in the Gulf of Alaska, researchers speculate that allonursing results from two things: the inability of inexperienced mothers to recognize their own offspring and the ability of young pups to steal milk from other mothers while they're sleeping.[163] When alien pups attempt to allonurse from an experienced, wide-awake mother, they are aggressively removed and even thrown or bitten. This suggests that there is a considerable cost to allonursing.

A long-term study of female African lions (*Panthera leo*) and their cubs in the Serengeti National Park[164] revealed that extensive allonursing occurs in this species for inclusive fitness benefits and to evacuate excess milk. Although mothers preferentially nursed their own offspring, they were also likely to allonurse offspring of close kin such as sisters and cousins. Female lions living in crèches with first-order female relatives nursed each other's offspring equally. In addition, mothers with smaller litters (and presumably more milk to spare) were found to allonurse more frequently, as were those whose own cubs had grown larger and therefore had lower milk demands.

So where do humans fit in? Allonursing, or wet nursing, was commonplace until the mid–twentieth century, and it is still widespread in some societies. Since a human baby is utterly incapable of assessing its own nutritional status, much less doing anything about it, milk stealing practices and attempts at allonursing observed in other species are nonexistent in *Homo sapiens*. For us, allonursing is entirely the decision of the mother and is most often due to difficulties with milk production. In our society, we are quite unlikely to allonurse anyone else's children, whether they are close kin or not. Women who have excess breast milk can donate to milk banks at local hospitals, so I suppose that excess milk production in our species can indirectly result in allonursing of unrelated offspring in a very limited form. For some women, the need to find an alternative to breastfeeding stems from a rather unnatural human practice: implantation of silicone into the breasts for aesthetic purposes. I'm certain that humans are alone in the animal kingdom on that one.

Pick a Favourite!

As parents, most of us vow to keep things equal between our offspring. In theory there should be an equal amount of time invested in Suzie's piano lessons as in Bobby's tennis practice. But in reality these things are never quite equal, are they? Bobby's tennis practice is across town, making the investment in his achievements greater than when Suzie walks over to the lady down the street for her piano lesson. Next, the inevitable—Suzie hates piano but Bobby looks as though he's heading straight to the Olympic trials with his perfected backswing. Resource allocation between offspring is suddenly a far cry from fifty-fifty. We can say that we invest equally in our offspring, but this is an almost impossible task in the animal kingdom, including *Homo sapiens.*

One complicating factor is how to handle offspring with different fathers. Perhaps Bobby was destined to be a champion based on the fact that his biological father is an admirable member of society and a champion athlete, but poor little Suzie was the outcome of a one-night stand after a few too many cocktails with a character whose name you'd rather erase from memory. Life history theory predicts that maternal investment in offspring should reflect the likelihood of the offspring to contribute to the fitness of the mother. This could be bad news for Suzie.

Many species in the animal kingdom mate with different partners in different mating seasons. Based on a number of environmental, physical, and random factors, a female may not always end up with the mate she most desires. However, offspring are the inevitable result, and so an interesting conundrum presents itself: invest equally in all offspring despite the suboptimal paternity of some? Biologically speaking, the answer should be a resounding no!

According to the differential allocation hypothesis,[165] females of many species do not contribute equally to the health and well-being of offspring sired by fathers of different quality. In experimental mating research on mallard ducks (*Anas platyrhynchos*), researchers paired individual females with both high-ranking and low-ranking males in order to investigate whether they would invest differently in their offspring based on the identity of the father.[166] Egg size is entirely determined by the female, and is a critical trait influencing fitness in birds: larger eggs produce larger chicks that have an increased chance of surviving to adulthood. Sure enough, it was demonstrated that eggs are significantly larger when females mate with a high-ranking male. Invest equally in all offspring? I think not.

Blue-footed booby (*Sula nebouxii*) males have a clear signal to show females that they are highly fit and ready to sire the next generation: their feet. Female boobies prefer males with bright blue-green feet and will actively discriminate against males without this characteristic. Once copulation has occurred, the female generally lays two eggs, the second approximately four days after the first. Both parents incubate the clutch and raise the young together until they are ready to leave the nest. In an experiment designed to assess the differential allocation hypothesis, researchers painted the feet of the male partner to a dull, unattractive, colour just after the first egg was laid.[167] As a result of this change in the male's foot morphology, the second egg laid by the female partner had a lower volume as well as a lower hormonal content. So perceptive is the female blue-footed booby that within a few days after her mate's quality is lowered, she reacts accordingly in her egg investment.

The effects of partner attractiveness do not end at the physiological stage of egg-laying. It has been shown for Zebra finches (*Taeniopygia guttata*) that the amount of postnatal parental care by both mothers and fathers varies with partner quality. Parental care activities like nest maintenance, watching over young offspring, and brooding, feeding, and grooming them are generally performed by both Zebra finch parents, but levels of all these activities are lower when a partner has an unattractive mate.[168] It's as if parents are reluctant to place all their reproductive resources into offspring that may have mediocre genes. Tough luck, kid.

I suppose most human parents would argue against applying the differential allocation hypothesis in our species, and in many cases this would be correct. Most *Homo sapiens* parents do an admirable job of investing equally in their offspring. Many parents raise children who are not only born of 'unattractive' parents but of parents who aren't biologically connected to them. Adoption, foster parenting, and stepparenting are commonplace among humans. Despite the fact that at times Bobby's tennis talents may overshadow the accomplishments of his sister, as a human offspring Suzie is likely to be well provided for. Perhaps the animal kingdom could learn a thing or two about the importance of the nurture part of the nature vs. nurture argument.

Another Girl?

'You've got the million dollar family!' were words I heard over and over after I gave birth to my second child—a daughter, after first having had a son. How perfectly random that I gave birth to a boy and then to a girl! I felt like a small-scale exhibit of Fisher's (1930)[169] theory of sex allocation whereby the ratio of males to females should remain constant at around fifty-fifty. Say there's a skew in the sex ratio of a population and it has more males than females. It makes sense for parents to invest in females, the rarer sex, in order to maximize their chances of future reproductive success. A daughter is virtually guaranteed to reproduce with a top choice from a population of hopeful bachelors, which is good news for the biological fitness of her parents. However, if the entire population starts investing in the rarer sex, the sex ratio will soon swing back to an approximately even split. The long-term result of this kind of strategy is strong selection for an equal investment in each of the two sexes. If there are too many of one sex, selection becomes stronger for the other to correct the disparity.

When Fisher coined his theory based on occurrences in the animal kingdom, I'm certain that he wasn't thinking about the future of human biology. He could never have envisioned the possibility that a legislatively-imposed bias in human child-rearing in one of the most densely populated countries of the world would result in a gender imbalance unprecedented in the animal kingdom. It isn't news that the one-child policy introduced in China in 1980 flies in the face of sex allocation theory by selectively favouring males. Reasons for keeping male children over female children were mostly due to a firmly entrenched tradition of parents wanting someone to support them when they get old and expecting that to be the son. A female child would go to the family of her future husband, and so there would likely be no help for her parents. To put it in biological terms, girls were much too costly to produce.

Humans are not alone in the animal kingdom when it comes to sex selection of offspring. Parents of diverse animal species are expected to bias the sex ratio of their offspring, according to Fisher's theory, in order to maximize their own biological fitness based on the ambient physiological and environmental conditions. Factors like territory quality, maternal/paternal condition and competitive ability,[170] and kin interactions in group-living species can all influence whether the sex ratio at a particular time is biased towards males or females. In Zebra finches (*Taeniopygia guttata*), sex selection of offspring occurs both primarily (at the time of egg production) and secondarily (adult females kill hatchlings of the unwanted sex).[171] For these organisms the sex ratio adjustment is determined by resource availability. When food is scarce, mostly male offspring are produced, because a lower body mass at the time of fledging does not affect their ability to produce young when they are adults themselves. The same is not true for females: those with a low fledgling body mass are less fecund as adults and therefore have less reproductive success. It makes sense that the finches place all their eggs in the male basket (pun intended) when resources are scarce.

Hermaphrodites present an interesting opportunity to investigate biases in sex allocation, since the same individual can produce both male and female gametes. The local mate competition theory[172] predicts that a female-biased sex ratio should be observed when males compete for chances at fertilization and females do not. However, as the number of females rises, increasing the potential for competition between them, more males should be produced instead. Researchers investigating the investment of hermaphroditic polycheate worms (*Ophryotrocha diadema*) into male and female gametes in both large (high-competition) and small (low-competition) groups found the expected differences.[173] When group size was large, more male gametes were produced, and when group size was small, more female gametes were produced. These organisms clearly modify their sex allocation in response to the level of local competition.

As these examples from diverse phyla show, adjustments in the sex ratio are common in the animal kingdom. However, the majority of species, including finches and polycheates, undergo several reproductive bouts over their life spans, allowing temporary biases in the sex ratio to even out as conditions change. No other species is limited by a one-clutch statute imposed on it by some omnipotent ruling body. What will happen to the nearly 30 million 'extra' males in China who will reach marrying age by 2020? Most will likely die without passing on their genetic blueprints. Family lineages will die off, and entire heritages will be lost. This is a perfect example of how humans ignore the fact that, underneath our complexities, we are still biological beings. To impose rules on the reproduction part of the 'survive and reproduce' mantra may well be crossing a line that was not meant to be crossed.

Sibling Rivalry at Its Worst

Firstborn offspring have it pretty good. Mom is generally healthy and well-rested, since she had no other children to look after during her pregnancy, and the newly emerged microhuman has the exclusive attention of both parents and quite likely some grandparents as well. The firstborn doesn't have to share the resources of his/her parents as second and third children do. Being a mother of three children, I think about this often. I feel bad for my youngest child, who gets dragged around to preschool, music lessons, playdates, and the other endeavours of his older siblings. He inevitably gets his toys snatched away or his snacks devoured before he can get to them, and there's not a whole lot that can be done about it.

A major difference between humans and many other animals with respect to sibling competition is that, in other animals, firstborn offspring often attain independence before more offspring are born, allowing the parents to care for only one infant at a time. This isn't the case in *Homo sapiens*, whose offspring remain dependent too long for mothers to wait for independence before having more offspring—our physiology doesn't allow for 18 years between children. Instead, sacrifices are made and less parental care is given to all siblings. So why not just have one offspring and avoid the need to sacrifice any parental commitment? This could be risky. If all reproductive effort goes to a single offspring, and for whatever reason the offspring fails to reproduce, your biological fitness is doomed. The bet-hedging strategy[174] accounts for uncertain conditions in the future: there are clear advantages to having more offspring and allocating less parental care to each one.

Nonhuman mammals face a similar conundrum with repeat child-rearing. Fur seal and sea lion females rear a single offspring at a time and nurse it exclusively for a period of two or three years, but many females give birth to another offspring during the nursing phase. A long-term study of Galapagos fur seals (*Arctocephalus galapagoensis*) and sea lions (*Zalophus wollebaeki*) addressed the potential conflict between siblings competing for their mother's resources and also the conflict between mothers and offspring. Which offspring should she nurse? When should she choose the juvenile over the newborn and vice versa?[175]

Between-sibling conflict was found to be especially strong when resource levels were low and the older sibling was unable to forage effectively on its own, away from the supplementation of the mother's milk. During El Niño years, when there was low ocean productivity, between-sibling conflict often resulted in the continued nursing of the older juvenile and the death by starvation of the newborn. During periods of high resource availability, mothers aggressively defended their newborns against the juveniles' attempts to nurse. You might find yourself wondering: why should a female continue to reproduce when her newborn might starve or when her juvenile offspring isn't ready to stop nursing? The fluctuating conditions of the Galapagos make it impossible for the females to predict what will happen, making a bet-hedging strategy important. If food resources are plentiful when a new pup is born, it makes sense for the mother to defend the younger offspring from aggressive attacks by the older one: 'Get off my boob and go forage for yourself; there's plenty of food out there!' However, if food resources are scarce, the mother is expected to permit the intimidation—and starvation—of the younger sibling, since the older one has already received more of her resources and is more likely to survive to adulthood. Although the death of a pup is a high price to pay for a miscalculation in available resources, the mother still gets one healthy offspring when times are bad and has the potential for more when times are good.

The Galapagos study shows that bet-hedging is an effective reproductive strategy when environmental conditions and resource availability are unpredictable. Over the course of her reproductive lifetime, a female seal or sea lion can maximize her overall fitness by hedging her bets rather than by waiting for favourable conditions. Humans, on the other hand, have taken a lot of the unpredictability out of our ability to obtain resources, so our need for bet-hedging is reduced. While an El Niño year might affect the availability of fish in the fresh seafood section of the grocery store, it has no bearing on whether there will be enough food to go around. Although at times my youngest child has to fight a little harder for his share of resources, starvation is out of the question. Taking the nature out of foraging has its advantages.

He's Having a Baby!

It's a common topic of conversation among new (human) mothers: how our population would cease to exist if males had to bear the children. The physical costs, including a nine-month gestation followed by giving birth to a 6- to 10-lb. live infant, are extremely daunting and likely would not be readily undertaken even if the physiology of the human male permitted it. It doesn't end there. The duties of lactation and child care are generally responsibilities of the female parent, and such tasks involve a great deal of time and energy that could otherwise be spent creating more offspring to represent us in future generations. Human males seem to have it pretty good: biologically speaking, they, like the majority of males in the animal kingdom, contribute little more than genetic material to their offspring.

Although it may at first seem as though males get off easily when it comes to their ability to contribute to future generations, it's not all fun and games. Males almost universally compete with each other for sexual partners, and in many cases this leads to the evolution of elaborate physical structures, colouration, or behaviours.[176] In addition, the mere contribution of sperm to the reproductive tract of a female doesn't guarantee that a particular male's seed will be the successful competitor (see Artificial Insemination). Males and females both have difficulties when it comes to procreation, but the human female asserts that her male counterpart would not be willing to do the child-rearing in her place. Do any other females in the animal kingdom have it figured out better than we do?

Enter the family *Syngnathidae*, commonly referred to as pipefishes and seahorses. These fish are sex-role reversed, which means that males take on the pregnancy and childcare roles and females experience more intense competition for mates. Females deposit their eggs into a male's brooding pouch, and he is therefore guaranteed paternity once he fertilizes them.[177] The brood pouches found in different species are categorized from simple membranous egg compartments on the male's ventral side to fully enclosed brooding pouches with placenta-like structures.[178] The pregnant males take on the duties of osmoregulating the environment within the brood pouch, aerating the eggs, and providing nourishment. But females don't get off without some investment of their own. They are faced with an interesting conundrum when it comes to competing with each other for mates: egg production is still required, and it has a high energetic cost (unlike the metabolically cheap sperm production), so females don't have the same energetic freedom males do when it comes to producing sexual ornaments. If such ornaments are produced, there is a potential cost to fecundity, which can make the female less attractive to potential mates.[179] So what can a female sygnathid do to increase her chances to fill a male's brood pouch? Although they aren't as bizarre as structures seen on males in traditional sex roles, females do develop reproductive ornaments that are used both to attain mates and to deter other females. In addition to fussing with their ornaments, females attempt to ruin the reproductive efforts of other females. *Mating disruption* occurs when large females swim in between a male and female pair while they're mating (how rude!), effectively ending the transfer of eggs.[180] Large females are effective mating disrupters, and they have been shown to influence the behaviour of smaller females through intimidation. The mere presence of larger females can interfere with and substantially decrease reproduction in smaller ones.[181]

So when it comes to syngnathid fish, males are choosy and females are competitive. Perhaps it's just the human in me, but I feel that all that competing is undignified female behaviour. It simply isn't ladylike to disrupt a couple in the throes of passion, or even egg transfer. Although child-rearing is a difficult job (and I maintain that the human population would cease to exist should our sex roles be reversed), I think it's more empowering to be the one who chooses. No offence to all you males out there, but if we females are going to do all the work to care for your offspring, you'd better be prepared to compete for our affections.

When Times Are Tough: Two Working Parents

My neighbour and I had our daughters at approximately the same time. We spent a lot of time together during the first few months and had lots of playdates and tea time. We enjoyed the camaraderie and our daughters benefited from sharing each others' company. Then something terrible happened—my neighbour had to go back to work. Suddenly her tiny little child was without either of her parents for many long days at a time. I didn't see my neighbour for months as she and her husband began their new schedule of long work days, dealing with out-of-home day care, and caring for their daughter when they weren't at work. I don't think they saw much of each other, either, but this is reality. Times are tough, and in order to provide adequate resources for your family, the necessary steps must be taken. If this means leaving your newly hatched youngster with an alloparent while you go off to gather resources, then so be it.

During times of adversity, parents must find ways to provide the necessary resources for their offspring and for themselves, whether you're *Homo sapiens* or some other creature. Colonially breeding seabird parents must periodically leave their offspring unattended. Up to 98 per cent of such seabirds breed within densely populated nesting territories where food resources are not available.[182] Sustenance is collected from the ocean or elsewhere and brought back to the nesting area for consumption by parents and offspring. When chicks are left unattended by both parents (not common when food is easy to find) a form of alloparenting takes place. In common murre (*Uria aalge*) populations, nonbreeders and failed breeders in the colony have been shown to provide sufficient care to chicks that have been left alone to enable them to survive to fledging.[183] Some biologists speculate that the payoff for such alloparental care is a form of allopreening—grooming of the day care providers that is not entirely unlike the fees we pay to those who mind our own children. This kind of alloparental care seems to be a viable option for seabirds when conditions are favourable and parents don't spend too much time away from the nest simultaneously. However, when environmental conditions are adverse and food is scarce, the costs of leaving the offspring behind can escalate beyond the food benefits obtained by both parents foraging.

In a series of behavioural observations taken during a time of severe food shortage on the Isle of May in the UK, it was found that both parents of common murre offspring spent the majority of their time foraging away from the nest and their offspring.[184] More than 50 per cent of the young in the breeding colony were frequently left unattended. Did the alloparenting/allopreening tradeoff save the day for these unattended chicks? In a word, no. Instead of providing alloparental care, neighbouring conspecifics attacked unattended chicks, so aggressively and relentlessly that attacks accounted for 69 per cent of chick mortality during the season. It is thought that the abusive adults wanted to safeguard against adopting or mistakenly feeding the unattended chicks when their own offspring were also on the verge of starvation. The message is clear: when times are tough, take care of your own progeny first, not the poor little orphans who stumble into your territory.

Over 20 years of behavioural data for another colonial nesting seabird, Nazca boobies (*Sula granti*) in the Galapagos Islands, showed that up to 80 per cent of nonparental adults in a breeding colony engage in a mixture of behaviours with unattended chicks,[185] and these nonparental adult visits (NAVs) could take the form of aggressive or sexually abusive attacks. The chicks involved were generally at an age where their food demands were high enough to result in both parents foraging for some periods of the day, but young enough so as not to be able to repel the unwanted visits. Although aggressive attacks by NAVs on unattended chicks did not usually result in the death of the chick (as it did with the murres discussed above), lacerations made on the chicks' bodies left them vulnerable to death by blood-feeding ectoparasitic land birds. It is suggested that the abusive behaviour might result from the NAV attempting to eliminate future competition for mates (in same-sex attacks) or resources (in attacks on both sexes), but the fact that sexual behaviour towards the chicks was observed in 14.3 per cent of male NAV visits and 6.8 per cent of female NAV visits is as yet unexplained. Kind of makes you want to stretch your coupon savings a little farther, doesn't it?

Parents of many species are faced with a daunting tradeoff when it comes to providing for their young. When times are tough and dual foraging becomes a necessity, there is an unavoidable risk to the offspring, not only in terms of the reduction in emotional bonding between parent and offspring (of low importance to colonial seabirds; very important to humans), but in terms of suboptimal treatment of young when both parents are absent. Humans are far from being alone in the animal kingdom when it comes to making this difficult choice.

Irreconcilable Differences

Having known several married couples who have decided to call it quits, I can appreciate that humans don't always get it right when it comes to selection of a life partner. Divorce rates for our species in the western world are extremely high, leaving sociologists with a difficult puzzle to piece together. Despite the drama and divorce that we observe in our societies, there are many couples who stand the test of time, who stay near and dear for all their reproductive lives and beyond. As with humans, monogamous pair bonds are frequently observed in the animal kingdom and their long-term success varies just as ours does. Divorce in monogamous birds has been seen to range from as low as 0 per cent in waved albatrosses[186] *(Diomedea irrorata))* to almost 100 per cent in great blue herons[187] *(Ardea herodias)*. What makes monogamy work in some cases and not others?

It turns out that some behavioural characteristics are fairly good indicators of whether a pair will separate. A meta-analysis (a broad survey of research on a certain subject) of divorce rates in 158 bird species determined that certain attributes can predict the likelihood of divorce.[188] Birds that form part-time partnerships for a breeding season (rather than a continuous partnership), have a high mortality rate, live colonially, or are highly ornamented are more likely to experience divorce. There are clear, biologically relevant reasons for each situation.

First, part-time partnerships such as those in species with distinct breeding and nonbreeding seasons result in more frequent split-ups because of the increased effort it takes to find your partner when breeding season begins again. Second, a high mortality rate within a species is likely to result in behaviours that allow individuals to explore new partners in subsequent breeding seasons. Searching for a former partner wastes time and energy if the partner has died. Third, colonial species are more likely to divorce because multiple partners available for mating within a confined space isn't conducive to monogamy. Fourth, physical ornamentation in the animal kingdom generally indicates competition for mates and a strong case for sexual selection.[189] One must keep oneself well adorned in order to remain sexually successful, which doesn't bode well for a monogamous partnership (see Monogamy with a Twist).

Can such reasons for divorce be applied to the human species? First, divorce rates could be higher in partnerships where physical separation is frequent because of travel or long work hours, but in our species partners have no problem locating each other after a separation. Second, those in high-risk or life-threatening job situations leave their spouses at risk for becoming widowed, but this doesn't provide reasonable cause for divorce while the working partner is still alive. Third, humans aren't generally a colonial species. There are communal-living humans who could be at risk for divorce (or infidelity) because they live in close quarters with members of the opposite sex. Fourth, humans employ a high level of physical ornamentation, which can result in attention from members of the opposite sex. However, ornamentation in our species doesn't accurately signal health or fitness (see Are Those

Real?), and adornment may or may not be worn for the purpose of scoring a mate, regardless of the attention it garners.

The major difference between divorce in humans and divorce in other monogamous species is that irreconcilable differences in the animal kingdom roughly translates as *I think I can improve my overall fitness with a different partner.* A more fit partner could mean a partner who is found without a lot of energy expenditure, a partner who is more fit to reproduce simply by being alive, a partner who lives nearby and is available for additional procreation, or a partner who is appropriately physically ornamented, signalling a high degree of health and fitness. In the human world, 'irreconcilable differences' represents a host of issues having nothing to do with the biological fitness of either partner. In our species, nature is conspicuously absent from the divorce process.

Are You My Mommy?

My children don't look especially like my husband or me. However, unless there was some kind of weird switch at the hospital that I never found out about, I'm pretty certain they're mine. Now and then an expression crosses their little faces that re-iterates their status as my biological offspring, but as far as phenotype (outward appearance) goes, it's not obvious that I'm their mom. The diversity packed into our genetic code allows for a multitude of options when it comes to our appearance. There is a higher probability that one's own offspring will be closer in resemblance than the offspring of others, but there is certainly no guarantee. Members of our species, at least those of us who aren't identical twins or who don't look a lot like their relatives, are faced with the task of recognizing our own kin. For the most part, we do this easily through learned experiences such as living close together, communicating with close kin, and observing one another.

Another potential problem humans have mitigated through learned experiences is that close kin must be avoided in our search for appropriate partners for procreation. Mating with close relatives is, of course, a social no-no, and there are well-founded biological reasons for avoiding it. Could humans avoid such mating without actively learning our social context? No, we couldn't. Humans are remarkably nonbiological when it comes to kin recognition. We must rely solely on learned information to know with whom to associate in certain situations. So how do other species, without the social complexities of the *Homo sapiens*, manage to associate with their own biological families and avoid them when it comes to finding an appropriate mate?

In most cases, kin recognition is thought to be related to visual, olfactory, or acoustical signals. An action called *phenotype matching* occurs when individuals use a set of template cues from related individuals and apply the template to unfamiliar conspecifics to determine their relationship. Does this one look, smell, or sound like my immediate family? If so, we must be related, and we ought not to procreate. A special case of phenotype matching called the armpit effect[190] occurs when a self-template is compared with unfamiliar individuals—Does this one look, smell or sound like me?

How effective are these techniques? The recognition abilities of female three-spined sticklebacks (*Gasterosteus aculeatus*) are such that they can discern a potential mate from a brother using olfactory cues alone.[191] In laboratory experiments, females in large aquaria were provided with two computer simulations of courting males, one on each side of the aquarium. The visual simulations were identical, but on one side of the aquarium, water conditioned by a brother was added, while on the other side, water conditioned by an unrelated male was added. In this way researchers could determine whether the females used olfactory cues to discriminate kin from not-kin.

It was found that females adjusted their behaviour according to relatedness. Not only could they discern between brothers and nonbrothers, they could even make the distinction if they had had no previous contact with the brother. Researchers couldn't determine whether the recognition by the female was due to phenotype matching or to the armpit effect, but the result was clear: female sticklebacks can determine relatedness and use this information to maximize their biological fitness.

The urban cockroach (*Blattella germanica*) is a group-living species where several generations live together in a common area (and of course you're hoping that this area isn't close to where you are!). Although we *Homo sapiens* tend to show irreverence toward these organisms, in several ways they demonstrate a superior level of biological sophistication (see Why Humans Need a Food Pyramid). Once individuals have reached maturity, they don't disperse from a common living area, but subgroups in smaller areas begin to form. Having one's extended family living together in the same place has its advantages for group foraging and protection, but when it comes to finding that special someone, the situation is far from ideal.

A series of choice experiments on various developmental stages of these organisms revealed that cockroaches of all ages (even the larval nymphs!) chose to associate with siblings over nonsiblings.[192] Nymphs reared in isolation without contact with closely related individuals were able to discriminate effectively. The mechanism for this recognition is the assessment of cuticular hydrocarbons on the exoskeleton of each individual. This method would be like using a thumbprint, if humans could use thumbprints to recognize each other—cockroaches can discern whether an individual is closely related through antennal contact with the individual's cuticle. Siblings were preferred as living partners, but they weren't selected as mating partners, indicating that the choices made by individual cockroaches are context-dependent.

So when it comes to kin recognition on a biological level, humans fail miserably. Even if something like phenotype matching were possible in *Homo sapiens*, we mask our natural scents with soaps and perfumes; we mask our natural looks with hair colour and makeup; and we cover our entire bodies with clothing that has nothing to do with our biological identity. There isn't much nature in human kin recognition, just learned behaviour. Time to give those cockroaches a little more respect!

Spoiling the Grandkids

Once in a while my mom takes my little ones for a Grandma sleepover. My husband and I try to ignore the junk food and video games that are in never-ending supply at grandma's house, because we get a chance at some one-on-one time together, a nice meal, or much-needed sleep! Grandparents are a wonderful thing. For the most part, we parents trust them to take care of our children when we need a break. First, they raised us, and since we turned out healthy and well enough to have offspring of our own, that's definitely a good sign. Second, they are 25 per cent related to our children, and that imparts a sense of biological closeness and responsibility. In any case, the fact remains that the grandparent gives up some resources (energy and time) in order to provide care to children who are not their own. Biologically speaking, why would they do this?

Since grandparents are often unable to reproduce at this time of life, the next best thing (biologically speaking) is to help rearing children who are close relatives. Human females are unique in the animal kingdom in that we reach reproductive senescence in the middle of our lifespan, leaving us in a position of biological irrelevance for a good portion of our lives. Menopause is thought to have evolved because middle-aged women can benefit more from contributing to the health and well-being of their grandchildren than from continuing to saddle themselves with offspring who take a greater physiological toll on an aging body—the grandmother hypothesis.[193] Most mammals experience a decline in reproductive output with aging, but they don't live long enough to experience reproductive senescence. However, in those who do, there are examples of senescent individuals, usually females, providing care to their grandchildren.

Having grandma around is beneficial to daughters and granddaughters of the Japanese macaque (*Macaca fuscata*) in several ways.[194] The reproductive pattern of females varies according to whether mom is there to help. Females with a living mother (both reproductive and post-reproductive) begin reproducing successfully at a younger age than females whose mother is deceased, which is thought to be due in part to protection (by her experienced mother) from advances by aggressive males. In addition, if a postreproductive grandmother is present, the daughter's interbirth interval—the length of time between births—is shorter. The additional care supplied by the grandmother, such as babysitting, protection, and support, provides the mother an opportunity to mate and concive sooner than she otherwise would.

The survival rate of newborn macaques up to one year old is affected by the status of the grandmother: experimental results show that 95 per cent survive when the grandmother is present and postreproductive, 89 per cent survive when the grandmother is present but still reproductive, and 85 per cent survive in absence of a grandmother.[195] It's clear that grandparents are a wonderful thing, but this second-generation help is short-lived, as most granny macaques don't survive much past the first few years of the life of their grandchildren.

Several long-lived and communally breeding bird species are appropriate candidates for studies of the relationships between grandparents and grandchildren. The Seychelles warbler (*Acrocephalus sechellensis*) is a communally breeding bird: one breeding pair breeds on and defends a territory where they raise their young. Subordinate nonbreeding individuals may also live on the territory and help care for the offspring. Among the helpers are older females who have been kicked out of their breeding role on a territory by their own daughters.[196] These newly subordinate grandmas can make the best of a suboptimal situation by providing provisional care to their daughters' offspring. It's not the warm, fuzzy grandmotherly care that we see in the human species, but to a warbler the survival of related offspring is better than having no offspring at all.

After all this, I feel pretty lucky that my mom is a willing participant in the raising of my children. Like the Japanese macaques, I've had fairly short intervals between childbirths. I'm sure that without the considerable support I get from my mother, I wouldn't have chosen this reproductive path. The extra help that I get permits me to gather my strength and energy so as to be a more effective parent to my children. I'm not convinced that the junk food, video games, and late nights my children enjoy with their grandma are supporting their health and well-being, but, all things considered, the break is totally worth it.

Postscript

Well, it's been quite a journey. Perhaps by now, from reading the examples in this book, you've come to your own conclusion about the nature in human nature. Despite the innumerable ways in which we humans tend to cover up our biology, the fact remains that we are simply another member of the animal kingdom. This book illustrates the unthinkable: in many ways the human animal is not as well equipped to survive and reproduce as many of our animal cousins. Clearly, a bigger brain doesn't mean increased biological fitness. In fact, one might ask a couple of crucial questions: Will the cranial capacity of *Homo sapiens* be our ultimate demise? Will we become too dependent upon unnatural interventions to live without them?

Probably, but until then I'll continue to enjoy the best of both worlds.

I'll concentrate on healthy, natural foods purchased in a ready-to-eat condition, but I might indulge in a fast-food cheeseburger now and then—when my children cannot see. I'll buy from local organic farms, and I'll never kill anything just for the sake of killing. I'll treat my children to the wonders of cuisines from around the globe.

I'll get my flu shot, and I'll eat chicken soup when struck down by a cold. I'll gratefully utilize the expertise of medical professionals, should I require it. I'll eat my purples and reds, and I'll try not to overmedicate when I have a fever. However, I won't quit my active use of mosquito repellent, and I certainly won't give up my mango-coconut shower gel.

I'll sleep safe and sound in my cozy, alarm-protected home and happily take my garbage out to the end of my driveway each Tuesday. I'll revel in the ease with which I can live my life without direct predators, although I'll remain alert so as not to meet my demise from a lack of attention. I'll do my part to be a good citizen, and I won't engage in physical confrontations with other members of our species (male or female), no matter how much it may be merited. I'll try to avoid getting caught up in trends that are detrimental to my health—but if I do, I'll keep in mind that moderation is key.

I won't give up my daily regimen of (nonsurgical) beauty interventions, but I'll admire members of the opposite sex (bad boys and good boys alike) only from afar. I'll continue to be sexually monogamous (and to enjoy the irrelevant existence of the female orgasm) with my wonderful husband, although we will take steps to prevent any further spread of our genetic blueprints. Sometimes I will still have a headache. I will remain socially polygamous with my male and female friends and respect the lifestyle of those who are homosexual.

I'll hold the door open for as many pregnant gals as I can. I'll advocate for the health benefits of breastfeeding. I'll be a loving and giving parent, making sure that each of my offspring has sufficient resources. I'll try my best to distribute said resources between them in somewhat equal amounts. I'll even babysit for my neighbours and other friends who don't have the luxury of staying at home. My husband and I will continue to appreciate grandparent assistance whenever it comes our way.

Overall, I guess what I'm saying is that I'll try to be a little more aware of the nature both within me and around me. In this day and age, it's good to be a member of the human race. I'll continue to enjoy the ride with an unwavering respect for the many other species with whom we *Homo sapiens* share the planet, and I hope you will join me.

Endnotes

[1]Raubenheimer, D. and S.A. Jones. 2006. Nutritional imbalance in an extreme generalist omnivore: tolerance and recovery through complementary food selection. Animal Behavior 71: 1253-1262.

[2]Mayntz, D., Raubenheimer, D., M. Salomon, S. Toft, and S.J. Simson. 2005. Nutrient-specific foraging in invertebrate predators. Science 307:111-113.

[3]Krebs, J.R. 1978. Optimal foraging: Decision rules for predators. In: Krebs, J.R., Davies, N.B. eds. Behavioral Ecology. Oxford: Blackwell Scientific Publications, 23-63.

[4]Rosen, D.A.S. and A.W. Trites. 2000. Canadian Journal of Zoology 78: 1243-1250.

[5]Rosen, D. A. S. and Trites, A. W. 2004. Satiation and compensation for short-term changes in food quality and availability in young Steller sea lions (Eumetopias jubatus). Canadian Journal of Zoology 82: 1061-1069.

[6]Romano, M.D., J.F. Piatt, D.D. Roby. 2006. Journal of the Waterbird Society 29(4):407-524.

[7]Kitaysky, A. S. et al. 2006. A mechanistic link between chick diet and decline in seabirds? Proceedings of the Royal Society of London B 273: 445-450.

[8]Osterblom, H., Olsson, O., Blenckner, T. and Furness, R, W. 2008. Junk-food in marine ecosystems. Oikos 117: 967-977.

[9]Trites, A. W. et al. 2007. Bottom up forcing and the decline of the Steller sea lions (Eumetopias jubatus) in Alaska: assessing the ocean climate hypothesis. Fisheries Oceanography 16: 46-67.

[10]Tarnaud, L. and Yamagiwa, J. 2008. Age-dependent patterns of intensive observation on elders by free-ranging juvenile Japanese macaques (Macaca fuscata yakui) within foraging context on Yakushima. American Journal of Primatology 70: 1103-1113.

[11]Cadieu, N., Winterton, P. and Cadieu, J.C. 2008. Social transmission of food handling in the context of triadic interactions between adults and young canaries (Serinus canaria). Behavioral Ecology and Sociobiology 62: 795-804.

[12]Landys-Ciannelli, Piersma, T. and Jukema, J. 2003. Strategic size changes of internal organs and muscle tissue in the Bar-tailed Godwit during fat storage on a spring stopover site. Functional Ecology 17:151-159.

[13]Freed, L. A. 1981. Loss of mass in breeding wrens: stress or adaptation? Ecology 62: 1179–1186.

[14]Norberg, R.A. 1981. Temporary weight decrease in breeding birds may result in more fledged young. American Naturalist 118: 838–850.

[15]Niizuma, Y., Takahashi, A, Sasaki, N., Hayama, S., Tokita, N. and Watanuki, Y. 2001. Benefits of mass reduction for commuting flight with heavy food load in Leach's storm-petrel, *Oceanodroma leucorhoa*. Ecological Research 16: 197-203.

[16]Elliott, K.H., Jacobs, S.R., Ringrose, J., Gaston, A.J. and Davoren, G.K. 2008. Is mass loss in Brunnich's guillemots *Uria lomvia* an adaptation for improved flight performance or improved dive performance? Journal of Avian Biology 39: 619-628.

[17]Muller-Schwarze, D., H. Brashear, R. Kinnel, K.A. Hintz, A. Lioubomirov and C. Skibo. 2001. Food processing by animals: do beavers leach tree bark to improve palatability? Journal of Chemical Ecology 27(5): 1011-1028.

[18]Dearing, M.D. 1997. The manipulation of plant toxins by a food-hoarding herbivore, *Ochotona princeps*. Ecology 78(3): 774-781.

[19]Finn, J. Tregenza, T. and Norman, M. 2009. Preparing the perfect cuttlefish meal: complex prey handling by dolphins. Plos one 4: e4217.

[20]MacArthur, R.H. and Pianka, E.R. 1966. On optimal use of patchy environment. The American Naturalist 100: 603–609.

[21]Fontaine, C., Collin, C.L. and Dajoz, I. 2008. Generalist foraging of pollinators: diet expansion at high density. Journal of Ecology 96: 1002-1010.

[22]Alarcon, R., Waser, N.M. and Ollerton, J. 2008. Year to year variation in the topology of a plant-pollinator interaction network. Oikos 117: 1796-1807.

[23]Mahurin, E.J. and Freeberg, T.M. 2009. Chick-a-dee variation in Carolina chickadees and recruiting flockmates to food. Behavioral Ecology 20: 111-116.

[24]Hauser, M.D., Teixidor, P., Fields, L. and Flaherty, R. 1993. Food-elicited calls in chimpanzees: effects of food quantity and divisibility. Animal Behavior 45: 817-819.

[25]Di Bitetti, M.S. 2003. Food-associated calls and audience effects in tufted capuchin monkeys, *Cebus paella nigritus*. Animal Behavior 69: 911-919.

[26]Hauser, M.D. 1992. Costs of deception: cheaters are punished in rhesus monkeys (*Macaca mulatta*). Proceedings of the National Academy of Sciences, U.S.A. 89: 12137-12139.

[27]Fox, L.R. and Morrow, P.A. 1981. Specialization: Species Property or Local Phenomenon? Science 211: 887-893.

[28]Brooke-McEachern, M., Eagles-Smith, C.A., Efferson, C.M. and Van Vuren, D.H. 2006. Evidence for local specialization in a generalist mammalian herbivore, *Neotoma fuscipes*. Oikos 113: 440-448.

[29]Anderson, R.C., Wood, J.B., and Mather, J.A. 2008. *Octopus vulgaris* in the Caribbean is a specializing generalist. Marine Ecology Progress Series 371: 199-202.

[30]Way, M.J. 1963. Mutualism between ants and honeydew-producing homoptera. Annual Review of Entomology 8: 307-344.

[31]Holldobler, B. and Wilson, E.O. 1990. Symbioses with other arthropods. Pages 471-529 in: *The Ants* Holldobler, B. and Wilson, E.O. eds. Belknap, Cambridge, Massachusetts, USA.

[32]Banks, C.J. and Nixon, H.L. 1958. Effects of the ant, Lasius niger L., on the feeding and excretion of the bean aphid Aphis fabae Scop. Journal of Experimental Biology 35: 703-711.

[33]Wimp, G.M. and Whitham, T.G. 2001. Biodiversity consequences of predation and host plant hybridization on an aphid-ant mutualism. Ecology 82: 440-452.

[34]Wimp, G.M. and Whitham, T.G. 2001. Biodiversity consequences of predation and host plant hybridization on an aphid-ant mutualism. Ecology 82: 440-452.

[35]Lounibos, L.P., Makhni, S., Alto, B.W. and Kesavaraju, B. 2008. Surplus killing by predatory larvae of *Corethrella appendiculata*: prepupal timing and site-specific attack on mosquito prey. Journal of Insect Behavior 21: 47-54.

[36]Corbet, P.S. and Griffiths, A. 1963. Observations on the aquatic stages of two species of *Toxorhynchites* (Diptera: Culicidae) in Uganda. Proceedings of the Royal Society of London 38: 125-135.

[37]Fincke, O.M. 1994. Population regulation of a tropical damselfly in the larval stage by food limitation, cannibalism, intraguild predation and habitat drying. Oecologia 100: 118-127.

[38]Huffman, M.A., Page, J.E., Sukhdeo, M., Gotoh, S., Kalunde, M., Chandrasiri, T. and Towers, G. 1996. Leaf-swallowing by chimpanzees: a behavioral adaptation for the control of strongyle nematode infections. International Journal of Primatology 17: 475-5003.

[39]Huffmann, M.A. and Caton, J.M. 2001. Self-induced increase of gut motility and the control of parasitic infections in wild chimpanzees. International Journal of Primatology 22: 329-346.

[40]Weldon, P.J. 200. Defensive anointing: extended chemical phenotype and unorthodox ecology. Chemoecology 14: 1-4.

[41]Weldon, P.J., Aldrich, J.R., Klun, J.A., Oliver, J.E. and Debbouin, M. 2003. Benzoquinones from millipedes deter mosquitoes and elicit self-anointing in capuchin monkeys (*Cebus* spp.). Naturwissenschaften 90: 301-304.

[42]Verderane, M.P., Falotico, T., Resende, B.D., Labruna, M.B., Izar, P. and Ottoni, E.B. 2007. Anting in a semifree-ranging group of *Cebus apella*. International Journal of Primatology 28: 47-53.

[43]Morrogh-Bernard, H.C. 2008. Fur-rubbing as a form of self-medication in *Pongo pygmaeus*. International Journal of Primatology 29: 1059-1064.

[44]Clark, L. 1991. The nest protection hypothesis: the adaptive use of plant secondary compounds by European starlings. In: Loye, J. E. and Zuk, M. (eds). Bird-parasite interactions: ecology, evolution, and behaviour. Oxford, Oxford University Press, pp. 205-221.

[45]Shutler, D. and Campbell, A.A. 2007. Experimental addition of greenery reduces flea loads in nests of a non-greenery using species, the tree swallow *Tachycineta bicolor*. Journal of Avian Biology 38: 7-12.

[46]Lafuma, L., Lambrechts, M.M. and Raymond, M. 2001. Aromatic plants in bird nests as a protection against blood-sucking flying insects? Behavioral Processes 56: 113-120.

[47]Castella, G., Chapuisat, M. and Christe, P. 2008. Prophylaxis with resin in wood ants. Animal Behavior 75: 1591-1596.

[48]Buehler, D.M., Piersma, T., Matson, K. and Tieleman, I.B. 2008. Seasonal redistribution of immune function in a migrant shorebird: annual-cycle effects override adjustments to thermal regime. The American Naturalist 172: 783-796.

[49]Povey, S., Cotter, S.C., Simpson, S.J., Lee, K.P. and Wilson, K. 2009. Can the protein costs of bacterial resistance be offset by altered feeding behaviour? Journal of Animal Ecology 78: 437-446.

[50]Lee, K.P., Cory, J.S., Wilson, K., Raubenheimer, D., and Simpson, S.J. 2006. Flexible diet choice offsets protein costs of pathogen resistance in a caterpillar. Proceedings of the Royal Society of London B Biological Sciences 273: 823-829.

[51]Schaefer, H.M., McGraw, K. and Catoni, C. 2008. Birds use fruit color as honest signal of dietary antioxidant rewards. Functional Ecology 22: 303-310.

[52]Roy, H.E., Steindraus, D.C., Eilenberg, J., Hajek, A.E., and Pell, J.K. 2006. Bizarre interactions and endgames: entomopathogenic fungi and their arthropod hosts. Annual Review of Entomology 51: 331-357.

[53]Ouedraogo, R.M., Goettel, M.S. and Brodeur, J. 2004. Behavioral thermoregulation in the migratory locust: a therapy to overcome fungal infection. Oecologia 138: 312-319.

[54]Foster, S.A. 1985. Wound healing: a possible role of cleaning stations. Copeia 1985: 875-880.

[55]Bshary, R. and Grutter, A.S. 2002. Parasite distribution on client reef fish determines cleaner fish foraging patterns. Marine Ecology Progress Series 235: 217-222.

[56]Grutter, A.S. 1999. Cleaner fish really do clean. Nature 398: 672-673.

[57]Grutter, A.S. 1999. Cleaner fish really do clean. Nature 398: 672-673.

[59]French, S.S., DeNardo, D.F. and Moore, M.C. 2007. Trade-offs between the reproductive and immune systems: facultative responses to resources or obligate responses to reproduction? The American Naturalist 170: 79-89.

[60]Birkeland, K. and Jakobsen, P.J. 1997. Salmon lice, *Lepeophtheirus salmonis*, infestation as a causal agent of premature return to rivers and estuaries by sea trout, *Salmo trutta*, juveniles. Environmental Biology of Fishes 49: 129-137.

[61]Finstad, B., Bjorn, P.A. and Nilsen, S.T. 1995. Survival of salmon lice, *Lepeophtheirus salmonis* Kroyer, on Arctic charr, *Salvelinus alphinus* (L.), in fresh water. Aquaculture Research 26: 791-795.

[62]Webster, S.J., Dill, L.M. and Butterworth, K. 2007. The effect of sea lice infestation on the salinity preference and energetic expenditure of juvenile pink salmon (*Oncorhynchus gorbuscha*). Canadian Journal of Fisheries and Aquatic Sciences 64: 672-680.

[63]Shih, H.T., Mok, H.K. and Chang, H.W. 2005. Chimney building by male Uca formosensis after pairing: a new hypothesis for chimney function. Zoological Studies 44: 242-251.

[64]Slatyer, R.A., Fok, E.S., Hocking, R. and Blackwell, P.R.Y. 2008. Why do fiddler crabs build chimneys? Biology Letters 4: 616-618.

[65]Schuetz, J.G. 2004. Common waxbills use carnivore scat to reduce the risk of nest predation. Behavioral Ecology 16: 133-137.

[66]Chase, I.D. 1991. Vacancy chains. Annual Review of Sociology 17: 133-154.

[67]Lewis, S.M. and Rotjan, R.D. 2009. Vacancy chains provide aggregate benefits of *Coenobita clypeatus* hermit crabs. Ethology 115: 356-365.

[68]Chase, I.D., Weissburg, J. and Dewitt, T.H. 1988. The vacancy chain process: a new mechanism of resource distribution in animals with application to hermit crabs. Animal Behavior 36: 1265-1274.

[69]Rotjan, R.D., Blum J. and Lewis, S.M. 2004. Shell choice in Pagurus longicarpus hermit crabs: does predation threat influence shell selection behavior? Behavioral Ecology and Sociobiology 56: 171-176.

[70]Wickler, W. 1968. Mimicry in plants and animals. New York: McGraw Hill.

[71]Cheney, K.L. 2008. The role of avoidance learning in an aggressive mimicry system. Behavioral Ecology 19: 583-588.

[72]Cote, I.M. 2004. Distance-dependent costs and benefits of aggressive mimicry in a cleaning symbiosis. Proceedings of the Royal Society of London, B Biological Sciences. 271: 2627-2630.

[73]Dawkins, R. and Krebs, J.R. 1979. Arms races between and within species. Proceedings of the Royal Society of London, B Biology. 205: 489-511.

[74]Clucas, B., Owings, D.H. and Rowe, M.P. 2008. Donning your enemy's cloak: ground squirrels exploit rattlesnake scent to reduce predation risk. Proceedings of the Royal Society of London, B Biological Sciences 275: 847-852.

[75]Montgomerie, R., Lyon, B. and Holder, K. 2001. Dirty ptarmigan: behavioral modification of conspicuous male plumage. Behavioral Ecology 12: 429-438.

[76]Berke, S.K. and Woodin, S.A. 2008. Energetic costs, ontogenetic shifts and sexual dimorphism in spider crab decoration. Functional Ecology 22: 1125-1133.

[77]Dukas, R. and Kamil, A.C. 2000. The cost of limited attention in blue jays. Behavioral Ecology 11: 502-506.

[78]Wilcox, R.S., Jackson, R.R. and Gentile, K. 1996. Spider web smokescreens: spider trickster uses background noise to mask stalking movements. Animal Behavior 51: 313-326.

[79]Dornhaus, A. 2008. Specialization does not predict individual efficiency in an ant. PloS Biology 6: 2368-2375.

[80]Dornhaus, A., Holley, J., Pook, V.G., Worswick, G. and Franks N.R. 2008. Why do not all workers work? Colony size and workload during emigrations in the ant Temnothroax albipennis. Behavioral Ecology and Sociobiology 63: 43-51.

[81]Hölldobler B., and Wilson E.O. 1990. The ants. Harvard University Press, Cambridge, MA

[82]Draud, M., Macias-Ordonez, R., Verga, J. and Itzkowitz, M. 2004. Female and male Texas cichlids (*Herichthys cyanoguttatum*) do not fight by the same rules. Behavioral Ecology 15: 102-108.

[83]Robinson, S.K. 1985. Fighting and assessment in the yellow-rumped cacique (*Cacicus cela*). Behavioral Ecology and Sociobiology 18: 39-44.

[84]Briffa, M. and Dallaway, D. 2007. Inter-sexual contests in the hermit crab *Pagurus bernhardus*: females fight harder but males win more encounters. Behavioral Ecology and Sociobiology 61: 1781-1787.

[85]Briffa, M. and Dallaway, D. 2007. Inter-sexual contests in the hermit crab *Pagurus bernhardus*: females fight harder but males win more encounters. Behavioral Ecology and Sociobiology 61: 1781-1787.

[86]Figler, M.H., Blank, G.S. and Peeke, H.V. 2005. Shelter competition between resident male red swamp crayfish *Procambarus clarkii* (Girard) and conspecific intruders varying by sex and reproductive status. Marine and Freshwater Behavior and Physiology 38: 237-248.

[87]Valone, T.J. 2007. From eavesdropping on performance to copying the behavior of others: a review of public information use. Behavioral Ecology and Sociobiology 62: 1-14.

[88]MacArthur, R.H., MacArthur, J.W. and Preer, J. 1962. On bird species diversity II. Prediction of bird censuses from habitat measurements. American Naturalist 96: 167-174.

[89]Betts, M.G., Hadley, A.S., Rodenhouse, N. and Nocera, J.J. 2008. Social information trumps vegetation structure in breeding-site selection by a migrant songbird. Proceedings of the Royal Society of London, B Biological Sciences 275: 2257-2263.

[90]Sato, Y. and Saito, Y. 2008. Evolutionary view of waste-management behavior using volatile chemical cues in social spider mites. Journal of Ethology 26: 267-272.

[91]Hart, A.G. and Ratnieks, F.L. 2002. Waste management in the leaf-cutting ant *Atta colombica*. Behavioral Ecology 13: 224-231.

[92]Hart, A.G. and Ratnieks, F.L. 2002. Waste management in the leaf-cutting ant *Atta colombica*. Behavioral Ecology 13: 224-231.

[93]Massey, A. and J.G. Vandenbergh. 1980. Puberty Delay by a Urinary Cue from Female House Mice in Feral Populations. Science 209: 821-822.

[94]Lombardi, J.R. and Vandenbergh, J.G. 1977. Pheromonally induced sexual maturation in females: regulation by the social environment of the male. Science 196: 545-546.

[95]Maggioncalda, A.N., N.M. Czekala, and R.M. Sapolsky. 2002. Male Orangutan Subadulthood: A New Twist on the Relationship Between Chronic Stress and Developmental Arrest. American Journal of Physical Anthropology 118: 25-32.

[96]Galdikas, B. 1985. Subadult male orangutan sociality and reproductive behaviour at Tanjung Putting. American Journal of Primatology 8: 87-99.

[97]Gadagkar R. 2003. Is the peacock merely beautiful or also honest? Current Science 85:1012–1020

[98]Budden, A.E. and Dickinson, J.L. 2009. Signals of quality and age: the information content of multiple plumage ornaments in male western bluebirds Sialia mexicana. Journal of Avian Biology 40: 18-27.

[99]Hunt, S., Bennett, A.T.D., Cuthill, I.C. and Griffiths, R. 1998. Blue tits are ultraviolet tits. Proceedings of the Royal Society of London, B. 265: 451-455.

[100]Bonato, M., Evans, M.R. and Cherry, M.I. 2009. Investment in eggs is influenced by male coloration in the ostrich Struthio camelus. Animal Behavior 77: 1027-1032.

[101]Hampton, K.J., Hughes, K.A. and Houde, A.E. 2009. The allure of the distinctive: reduced sexual responsiveness of female guppies to 'redundant' male color patterns. Ethology 115: 475-481.

[102]Andersson, M. 1994. Sexual Selection. Princeton University Press, Princeton, New Jersey.

[103]Guevara-Fiore, P., Skinner, A. and Watt, P.J. 2009. Do male guppies distinguish virgin females from recently mated ones? Animal Behavior 77: 425-431.

[104]Durgin, W.S., Martin, K.E., Watkins, H.R. and Mathews, L.M. 2008. Distance communication of sexual status in the crayfish *Orconectes quinebaugensis*: female sexual history mediates male and female behavior. Journal of Chemical Ecology 34: 702-707.

[105]Iyengar, V.K. 2009. Experience counts: females favor multiply mated males over chemically endowed virgins in a moth (*Utetheisa ornatrix*). Behavioral Ecology and Sociobiology 63: 847-855.

[106]Gilby, I.C. 2006. Meat sharing among Gombe chimpanzees: harassment and reciprocal exchange. Animal Behavior 71: 953-963.

[107]van Noordwijk, M.A. and van Schaik, C.P. 2009. Intersexual food transfer among orangutans: do females test males for coercive tendency? Behavioral Ecology and Sociobiology 63: 883-890.

[108]Jaeger, R.G., Gillette, J.R., Cooper, R.C. 2002. Sexual coercion in a terrestrial salamander: males punish socially polyandrous female partners. Animal Behavior 63: 871-877.

[109]Liebgold, E.B., Cabe, P.R., Jaeger, R.G., and Leberg, P.L. 2006. Multiple paternity in a salamander with socially monogamous behavior. Molecular Ecology 15: 4153-4160.

[110]Bussiere, L.F., Basit, H.A., Gwynne, D.T. 2004. Preferred males are not always good providers: female choice and male investment in tree crickets. Behavioral Ecology 16: 223-231.

[111]Bussiere, L.F., Basit, H.A., Gwynne, D.T. 2004. Preferred males are not always good providers: female choice and male investment in tree crickets. Behavioral Ecology 16: 223-231.

[112]Madden, JR. and A. Balmford. 2004. Spotted bowerbirds Chlamydera maculata do not prefer rare or costly bower decorations. Behavioral Ecology and Sociobiology 55: 589-595.

[113]Doucet, S.M. and R. Montgomerie. 2003. Multiple sexual ornaments in satin bowerbirds: ultraviolet plumage and bowers signal different aspects of male quality. Behavioral Ecology 14(4): 503-509.

[114]Wojcieszek, J.M., Nicholls, J.A. and Goldizen, A.W. 2007. Stealing behavior and maintenance of a visual display in the sating bowerbird. Behavioral Ecology 18: 689-695.

[115]Borgia, G. and M.A. Gore. 1986. Feather stealing in the satin bowerbird (Ptilonorhynchus violaceus)- male competition and the quality of display. Animal Behavior 34: 727-738.

[116]Borgia, G. 1993. The cost of display in the non-resource based mating system of the satin bowerbird. American Naturalist 141: 729-743.

[117]Ophir, A.G., and Galef, B.G. 2003. Female Japanese quail that 'eavesdrop' on fighting males prefer losers to winners. Animal Behavior 66: 399-407.

[118]Ophir, A.G., and Galef, B.G. 2003. Female Japanese quail that 'eavesdrop' on fighting males prefer losers to winners. Animal Behavior 66: 399-407.

[119]Wong, B.B. 2004. Superior fighters make mediocre fathers in the Pacific blue-eye fish. Animal Behavior: 67: 583-590.

[120]Steiner, S., J.L.M. Steidle and J. Ruthner. 2005. Female sex pheromone in immature insect males- a case of pre-emergence chemical mimicry? Behavioral Ecology and Sociobiology 55:111-120

[121]Trail, P.W. 2009. Why should lek breeders be monomorphic? Evolution 44: 1837-1852.

[122]Dugatkin, L.A. and Druen, M. 2007. Mother-offspring correlation and mate-choice copying behavior in guppies. Ethology Ecology and Evolution 19: 137-144.

[123]Hill, S.E. and Ryan, M.J. 2006. The role of model female quality in the mate- choice copying behavior of sailfin mollies. Biology Letters 2: 203-205.

[124]Amlacher, J. and Dugatkin, L.A. 2005. Preference for older over younger models during mate-choice copying in young guppies. Ethology Ecology and Evolution 17: 161-169.

[125]Witte, K. and Massmann, R. 2003. Female sailfin mollies, Poecilia latipinna, remember males and copy the choice of others after 1 day. Animal Behavior 65: 1151-1159.

[126]Higham, J.P., Ross, C., Warren, Y., Heistermann, M., and MacLarnon, A.M. 2007. Reduced reproductive function in wild baboons (*Papio hamadryas anubis*) related to natural consumption of the African black plum (*Vitex doniana*). Hormones and Behavior 52: 384-390.

[127]Neiman, M. and Lively, C.M. 2005. Male New Zealand mud snails (*Potamopyrgus antipodarum*) persist in copulating with asexual and parasitically castrated females. American Midland Naturalist 154: 88-96.

[128]Neiman, M. and Lively, C.M. 2005. Male New Zealand mud snails (*Potamopyrgus antipodarum*) persist in copulating with asexual and parasitically castrated females. American Midland Naturalist 154: 88-96.

[129]Snell, T.W. and Childress, M. 1987. Aging and loss of fertility in male and female *Brachionus plicatilis* (Rotifera). International Journal of Invertebrate Reproduction and Development 12: 103-110.

[130]Gomez, A. and Serra, M. 1996. Mate choice in male *Brachionus pllicatilis* Rotifers. Functional Ecology 10: 681-687.

[131]Takami, Y., Sasabe, M., Nagata,N., and Sota, T. 2008. Dual function of seminal substances for mate guarding in a ground beetle. Behavioral Ecology 19: 1173-1178.

[132]Uhl, G. and Busch, M. 2009. Securing paternity: mating plugs in the dwarf spider *Oedothorax retusus* (*Araneae: Erigoniae*). Biological Journal of the Linnean Society 96: 574-583.

[133]Segoli, M., Lubin, Y., and Harari, A.R. 2008. Frequency and consequences of damage to male copulatory organs in a widow spider. The Journal of Arachnology 36: 533-537.

[134]Nessler, S.H., Uhl, G. and Schneider, J.M. 2009. Sexual cannibalism facilitates genital damage in *Argiope lobata* (*Araneae: Araneidae*). Behavioral Ecology and Sociobiology 63: 355-362.

[135]Rowe, L., Arnqvist, G., Sih, A. and Krupa, J.J. 1994. Sexual conflict and the evolutionary ecology of mating patterns: water striders as a model system. Trends in Ecology and Evolution 9:289-293.

[136]Arnqvist, G. 1988. Mate guarding and sperm displacement in the water strider *Gerris lateralis* Schumm. (Heteroptera: Gerridae). Freshwater Biology 19: 269-274.

[137]Arnqvist, G. and Rowe, L. 2002. Correlated evolution of male and female morphologies in water striders. Evolution 56: 936-947.

[138]Karino, K., and Urano, Y. 2008. The relative importance of orange spot coloration and total length of males in female guppy mate preference. Environmental Biology of Fishes 83: 397-405.

[139]Pilastro, A., Simonato, M., Bisazza, A. and Evans, J.P. 2004. Cryptic female preference for colorful males in guppies. Evolution 58: 665-669.

[140]Pizzari, T. and Birkhead, T.R. 2000. Female feral fowl eject sperm of subdominant males. Nature 405: 787-789.

[141]Ursprung, C., den Hollander, M., and Gwynne, D.T. 2009. Female seed beetles, *Callosobruchus maculatus*, remate for male-supplied water rather than ejaculate nutrition. Behavioral Ecology and Sociobiology 63: 781-788.

[142]Gwynne, D. 1993. Food quality controls sexual selection in mormon crickets by altering male mating investment. Ecology 74: 1406-1413.

[143]Takakura, K. 2004. The nutritional contribution of males affects the feeding behavior and spatial distribution of females in a bruchid beetle, *Bruchidius dorsalis*. Journal of Ethology 22:37-42.

[144]Edvardsson, M. 2007. Female Callosobruchus maculatus mate when they are thirsty: resource-rich ejaculates as mating effort in a beetle. Animal Behavior 74: 183-188.

[145]Darwin, C. 1871.The descent of man, and selection in relation to sex. Murray, London

[146]Balenger, S.L., Johnson, L.S., and Masters, B.S. 2009. Sexual selection in a socially monogamous bird: male color predicts paternity success in the mountain bluebird, *Sialia currucoides*. Behavioral Ecology and Sociobiology 63: 403-411.

[147]Estep, L.K., Mays, H., Keyser, A.J., Ballentine, B.E., Hill G.E. 2005. Effects of breeding density and coloration on mate guarding and cuckoldry in blue grosbeaks (*Passerina caerulea*). Canadian Journal of Zoology 83:1143–1148.

[148]Gross, M. R. 1985. Disruptive selection for alternative life histories in salmon. Nature 313: 47–48.

[149]Wikelski, M.W. S. S. Steiger, B. Gall and K.N. Nelson. 2005. Sex, drugs, and mating role: testosterone-induced phenotype-switching in Galapagos marine iguanas. Behavioral Ecology 16(1): 260-268.

[150]Reinhardt, K. and Siva-Jothy, M.T. 2005. An advantage for young sperm in the house cricket *Acheta domesticus*. The American Naturalist 165: 718-723.

[151]Wikelski, M. and Baurle, S. 2009. Pre-copulatory ejaculation solves time constraints during copulations in marine iguanas. Proceedings of the Royal Society of London B Biological Sciences 263: 439-444.

[152]Watson P F. 1990. Artificial insemination and the preservation of semen. Pages 747–869 in *Marshall's Physiology of Reproduction*, Volume 2: Reproduction in the Male, Fourth Edition, edited by G E Lamming.New York: Churchill Livingstone.

[153]Bagemihl, B. 1999. Biological Exuberance: Animal Homosexuality and Natural Diversity. St Martin's Press, New York.

[154]Bagemihl, B. 1999. Biological Exuberance: Animal Homosexuality and Natural Diversity. St Martin's Press, New York.

[155]Levan, K.E., Fedina, T.Y. and Lewis, S.M. 2009. Testing multiple hypotheses for the maintenance of male homosexual copulatory behavior in flour beetles. Journal of Evolutionary Biology 22: 60-70.

[156]Bertran, J., Margalida, A. and Arroyo, B.E. 2009. Agonistic behavior and sexual conflict in atypical reproductive groups: the case of bearded vulture *Gypaetus barbatus* polyandrous trios. Ethology 115: 429-438.

[157]Zuk, M. and Bailey, N.W. 2008. Birds gone wild: same-sex parenting in albatross. Trends in Ecology and Evolution 23: 658-660.

[158]Cooper, W. E., Jr & Greenberg, N. 1992. Reptilian coloration and behavior. In: *Biology of the Reptilia: Hormones, Brain, and Behavior* (Ed. by C. Gans & D. Crews), pp. 298–422. Chicago: The University of Chicago Press.

[159]Watkins, G.G. 1997. Inter-sexual signaling and the functions of female coloration in the tropidurid lizard Microlophus occipitalis. Animal Behavior 53: 843-852.

[160]Gerald, M.S., Waitt, C. and Little, A.C. 2009. Pregnancy coloration in macaques may act as a warning signal to reduce antagonism by conspecifics. Behavioral Processes 80: 7-11.

[161]Gerald, M.S., Waitt, C. and Little, A.C. 2009. Pregnancy coloration in macaques may act as a warning signal to reduce antagonism by conspecifics. Behavioral Processes 80: 7-11.

[162]Roulin, A. 2002. Why do lactating females nurse alien offspring? A review of hypotheses and empirical evidence. Animal Behavior 63:201-208.

[163]Maniscalco, J.M., Harris, K.R., Atkinson, S. and Parker, P. 2007. Alloparenting in Steller sea lions (Eumetopias jubatus): correlations with misdirected care and other observations. Journal of Ethology 25: 125-131.S

[164]Pusey, A.E. and Packer, C. 1994. Non-offspring nursing in social carnivores: minimizing the costs. Behavioral Ecology 5:362-374.

[165]Burley N. 1986. Sexual selection for aesthetic traits in species with biparental care. American Naturalist 127(4):415–445

[166]Cunningham, E.J.A. and Russell, A.F. 2000. Egg investment is influenced by male attractiveness in the mallard. Nature 404: 74-77.

[167]Dentressangle, F., Boeck, L. and Torres, R. 2008. Maternal investment in eggs is affected by male feet color and breeding conditions in the blue-footed booby, Sula nebouxii. Behavioral Ecology and Sociobiology 62: 1899-1908.

[168]Burley N. 1988. The differential allocation hypothesis: an experimental test. American Naturalist 132:611–628.

[169]Fisher, R.A. 1930. The genetical theory of natural selection. Oxford: Oxford University Press.

[170]Trivers, R. L., and D. E. Willard. 1973. Natural selection of parental ability to vary the sex ratio of offspring. Science 179:90–92.

[171]Martins, T.L.F. 2004. Sex-specific growth rates in zebra finch nestlings: a possible mechanism for sex ratio adjustment. Behavioral Ecology 15(1): 174-180.

[172]Hamilton, W.D. 1967. Extraordinary sex ratios. Science 156: 477-488.

[173]Hermaphroditic worms have been documented to adjust their investment into male and female gamete production based on the number of potential sexual competitors that are present.[173]

[174]Mock, D.W. and Forbes, S. 1995. The evolution of parental optimism. Trends in Ecology and Evolution 7:409-413.

[175]Trillmich, F. and Wolf, J.B.W. 2008. Parent-offspring and sibling conflict in Galapagos fur seals and sea lions. Behavioral Ecology and Sociobiology 62:363-375.

[176]Darwin, C. (1871). "The descent of man and selection in relation to sex." Murray, London.

[177]Jones, A. G., G. Rosenqvist, A. Berglund, and J. C. Avise. 1999. The genetic mating system of a sex-role-reversed pipefish (Syngnathus typhle): a molecular inquiry. Behav. Ecol. Sociobiol. 46:357–365.

[178]Wilson, A.B., Ahnesjo, I., Vincent, A.C.J. and Meyer, A. 2003. The dynamics of male brooding, mating patterns, and sex roles in pipefishes and seahorses (family Syngnathidae). Evolution 57: 1374-1386.

[179]Berglund, A. and Rosenqvist, G. 2003. Sex role reversal in pipefish. Advances in the study of behavior 32: 131-167.

[180]Berglund, A. 1991. Egg competition in a sex role reversed pipefish: Subdominant females trade reproduction for growth. *Evolution* **45**, 770-774.

[181]Berglund, A. 1991. Egg competition in a sex role reversed pipefish: Subdominant females trade reproduction for growth. *Evolution* **45**, 770-774.

[182]Perrins, C. M. & Birkhead, T. R. 1983 Avian ecology. Glasgow, UK: Blackie.

[183]Birkhead, T. R. & Nettleship, D. N. 1984 Alloparental care in the common murre (*Uria aalge*). Can. J. Zool. 62, 2121–2124.

[184]Ashbrook, K., Wanless, S., Harris, M.P. and Hamer, K.C. 2008. Hitting the buffers: conspecific aggression undermines benefits of colonial breeding under adverse conditions. Biology letters 4: 630-633.

[185]Anderson, D. J., Porter, E. T. & Ferree, E. D. 2004 Nonbreeding Nazca boobies (*Sula granti*) show social and sexual interest in chicks: behavioural and ecological aspects. Behaviour 141, 959–977.

[186]Harris, M.P. 1973. The biology of the waved albatross *Diomedea irrorata* of Hood Island, Galapagos. Ibis 115: 483–510

[187]Simpson, K., Smith, J.N.M., Kelsall, J.P. 1987. Correlates and consequences of coloniality in great blue herons. Canadian Journal of Zoology 65: 572–577

[188]Jeschke, J.M. and Kokko, H. 2008. Mortality and other determinants of bird divorce rate. Behavioral Ecology and Sociobiology 63:1–9.

[189]Jeschke, J.M. and Kokko, H. 2008. Mortality and other determinants of bird divorce rate. Behavioral Ecology and Sociobiology 63:1–9.

[190]Dawkins, R. 1982. The extended phenotype. Freeman, San Francisco.

[191]Mehlis, M., Bakker, T.C.M. and Frommen, J.G. 2008. Smells like sib spirit: kin recognition in three-spined sticklebacks (*Gasterosteus aculeatus*) is mediated by olfactory cues. Animal Cognition 11: 643-650.

[192]Lihoreau, M. and Rivault, C. 2009. Kin recognition via cuticular hydrocarbons shapes cockroach social life. Behavioral Ecology 20: 46-53.

[193]Hill, K. & A.M. Hurtado, 1991. The evolution of premature reproductive senescence and menopause in human females. An evaluation of the grandmother hypothesis. *Human Nature* 2: 313–350.

[194]Pavelka, M.S., Fedigan, L.M. and Zohar, S. 2002. Availability and adaptive value of reproductive and postreproductive Japanese macaque mothers and grandmothers. Animal Behavior 64: 407-414.

[195]Pavelka, M.S., Fedigan, L.M. and Zohar, S. 2002. Availability and adaptive value of reproductive and postreproductive Japanese macaque mothers and grandmothers. Animal Behavior 64: 407-414.

[196]Richardson, D.S., Burke, T. and Komdeur, J. 2007. Grandparent helpers: the adaptive significance of older, postdominant helpers in the Seychelles warbler. Evolution 61: 2790-2800.